PUSHKAR JAIN

Tantalum Oxide Thin Films for Embedded Capacitors

PUSHKAR JAIN

Tantalum Oxide Thin Films for Embedded Capacitors

Using Pulsed DC Reactive Sputtering

VDM Verlag Dr. Müller

Impressum/Imprint (nur für Deutschland/ only for Germany)
Bibliografische Information der Deutschen Nationalbibliothek: Die Deutsche Nationalbibliothek
verzeichnet diese Publikation in der Deutschen Nationalbibliografie; detaillierte bibliografische
Daten sind im Internet über http://dnb.d-nb.de abrufbar.
Alle in diesem Buch genannten Marken und Produktnamen unterliegen warenzeichen-, marken-
oder patentrechtlichem Schutz bzw. sind Warenzeichen oder eingetragene Warenzeichen der
jeweiligen Inhaber. Die Wiedergabe von Marken, Produktnamen, Gebrauchsnamen,
Handelsnamen, Warenbezeichnungen u.s.w. in diesem Werk berechtigt auch ohne besondere
Kennzeichnung nicht zu der Annahme, dass solche Namen im Sinne der Warenzeichen- und
Markenschutzgesetzgebung als frei zu betrachten wären und daher von jedermann benutzt
werden dürften.

Coverbild: www.purestockx.com

Verlag: VDM Verlag Dr. Müller Aktiengesellschaft & Co. KG
Dudweiler Landstr. 125 a, 66123 Saarbrücken, Deutschland
Telefon +49 681 9100-698, Telefax +49 681 9100-988, Email: info@vdm-verlag.de
Zugl.: New York, Rensselaer Polytechnic Institute, PhD Dissertation, 2003

Herstellung in Deutschland:
Schaltungsdienst Lange o.H.G., Zehrensdorfer Str. 11, D-12277 Berlin
Books on Demand GmbH, Gutenbergring 53, D-22848 Norderstedt
Reha GmbH, Dudweiler Landstr. 99, D- 66123 Saarbrücken
ISBN: 978-3-639-09662-0

Imprint (only for USA, GB)
Bibliographic information published by the Deutsche Nationalbibliothek: The Deutsche
Nationalbibliothek lists this publication in the Deutsche Nationalbibliografie; detailed
bibliographic data are available in the Internet at http://dnb.d-nb.de.
Any brand names and product names mentioned in this book are subject to trademark, brand or
patent protection and are trademarks or registered trademarks of their respective holders. The use
of brand names, product names, common names, trade names, product descriptions etc. even
without
a particular marking in this works is in no way to be construed to mean that such names may be
regarded as unrestricted in respect of trademark and brand protection legislation and could thus
be used by anyone.

Cover image: www.purestockx.com

Publisher:
VDM Verlag Dr. Müller Aktiengesellschaft & Co. KG
Dudweiler Landstr. 125 a, 66123 Saarbrücken, Germany
Phone +49 681 9100-698, Fax +49 681 9100-988, Email: info@vdm-verlag.de

Produced in USA and UK by:
Lightning Source Inc., 1246 Heil Quaker Blvd., La Vergne, TN 37086, USA
Lightning Source UK Ltd., Chapter House, Pitfield, Kiln Farm, Milton Keynes, MK11 3LW, GB
BookSurge, 7290 B. Investment Drive, North Charleston, SC 29418, USA
ISBN: 978-3-639-09662-0

TABLE OF CONTENTS

ii

CHAPTER 4

BREAKDOWN VOLTAGE VERSUS CAPACITANCE DENSITY

CHAPTER 7

LIST OF TABLES

LIST OF FIGURES

ACKNOWLEDGMENT

First and foremost, I would like to acknowledge and thank my advisor Prof. E. J. Rymaszewski and Prof. T. S. Cale for their guidance and relentless support without which this thesis would not have been possible. Their knowledge, guidance, and support were crucial for the completion of this work. I would also like to acknowledge Prof. T. –M. Lu for guiding me like an advisor, and encouraging me to try out new ideas. I would like to thank my other committee members, Prof. R. J. Gutmann, and Prof. R. Kane for their guidance and valuable advice. The constructive criticism and guidance of my committee members helped me focus better on the thesis work.

I would also like to acknowledge Interconnect Focus Center and Center for Power Electronics Systems for providing me the financial support and research funding in various phases of this work.

Throughout this work, I have been helped by numerous other people in various capacities. I would like to thank Jasbir and Vinay, who helped me out tremendously throughout this work. Thanks are due to Anu and Tansel for sharing their immense experience in metal ion penetration and roughness evolution studies, respectively. I would also like thank Ravi and Jasbir for thoroughly reviewing this thesis, and providing valuable input in shaping this thesis manuscript. The clean room and the administrative staff at the Center for Integrated Electronics have been very helpful during my stay at RPI. I would like to thank Dana, Kelli, and Joelle for their help. My stay at Troy has been enjoyable because of the many great friends I made here. I would like to acknowledge Tuhin, Anju, Shom, Jasbir, Ravi, Amrit, Ashish, Rohit, Anu, Manisha, Tansel, Yongchai, Suchira, and Ed.

Last, but not the least, I would like to thank my family members for their love, support, and encouragement throughout the course of my academic career.

ABSTRACT

Embedded capacitor technology, where thin film capacitors are integrated at on-chip and/or off-chip levels, offers high packaging densities and improved electrical performance at potentially reduced costs of capacitor fabrication and integration. This research explores and establishes the leverages of using thin film embedded capacitors over currently used surface mount discrete capacitors. In particular, this thesis focuses on developing pulsed dc reactively sputtered tantalum oxide (Ta_2O_5) thin film capacitors to be integrated into established interconnect technologies of IC chips and packages.

A correlation between electrical breakdown field and dielectric constant, E_{BR} (MV/cm) = $\left(20/\sqrt{\varepsilon_r}\right)$ is empirically determined and used to establish a design space for breakdown voltage and capacitance density of planar capacitors, with film thickness and material dielectric constant as parameters. This design space sets the limits for "best one can achieve" (BOCA) breakdown voltages and capacitance densities using a particular dielectric. The validity of the developed design space is experimentally verified with Ta_2O_5 thin films over a wide range of film thickness (0.05 to 5.4 μm). Detractors causing the deviations from the BOCA breakdown voltages are identified and corrected experimentally. In particular, substrate cooling during deposition of "thicker" (> ~1 μm) films is required to deposit stoichiometric Ta_2O_5 exhibiting BOCA breakdown voltages, and "peak-to-valley" roughness at the metal/dielectric interfaces must be less than 10% of dielectric film thickness to avoid pre-mature breakdown of thin film capacitors.

An experimentally verified analytical model for pulsed dc reactive sputtering of Ta_2O_5 films is described and evaluated. The influences of important process variables, like oxygen flow rate and sputtering ion current, on the oxygen partial pressure in the chamber, deposition rate, film stoichiometry as well as film breakdown and leakage characteristics are predicted using this model. The experimentally established existence of multiple oxygen partial pressures at a given oxygen flow rate (hysteresis loop) is theoretically explained using steady state analysis. The experimental results suggest that in order to ensure the electrical reliability of Ta_2O_5 films, deposition should be done at oxygen flow rates more than that required to go beyond the hysteresis region in oxygen partial pressure versus flow rate curve.

High frequency test vehicles were designed and fabricated to evaluate the electrical performance of Ta_2O_5, SiO_2, and Si_3N_4 thin film capacitors over a wide range of frequencies (dc to 20 GHz). Ta_2O_5, SiO_2, and Si_3N_4 show no dispersion at least up to 20 GHz. The total inductance of power connect vias is determined to be less than 50 pH/μm of via, which is at least two orders of magnitude lower than most discrete capacitors along with connection leads (> 4 nH). Providing several vias in parallel further reduces the overall inductance of the connection paths in thin film capacitors. The shorter connections and elimination of connection leads in thin film decoupling capacitors make them highly suitable for integration into 2D and 3D structures. Thin film capacitors based on SiO_2, Si_3N_4, and Ta_2O_5 can provide capacitance densities comparable to the state-of-the-art surface mount discrete decoupling capacitors ($\sim$ 30 nF/cm^2). The leakage current densities below 10^{-6} A/cm^2 (at 0.5 MV/cm) are achieved with SiO_2, Si_3N_4 and Ta_2O_5 thin film dielectrics.

The extent of Cu diffusion/drift into Ta_2O_5 films is determined and compared with Al, Ta, and Ti at various biasing and temperature conditions using bias-temperature-stress (BTS) and triangular voltage sweep (TVS) techniques. No Cu diffusion was detected at 150 °C at least till 0.75 MV/cm. While, Cu diffuses/drifts into Ta_2O_5 at aggressive temperature conditions, i.e., at 300 °C was detected, a thin layer of Ti or Ta ($\sim$30 nm) acts as an excellent Cu diffusion barrier into Ta_2O_5.

CHAPTER 1

INTRODUCTION

The continued growth of microelectronics industry is hinged on ever increasing demands for higher functionality, speed, and portability of electronic products at reduced cost of manufacturing. This trend puts tremendous pressure on designers and manufacturers to pack more circuitry into smaller volumes. Continuous advances in photolithography have resulted in an enormous increase in the densities of semiconductor devices, circuits, and interconnections along with lower cost-per-function [1]. As integrated circuit (IC) technology is achieving faster speeds by shrinking feature sizes, the packaging at chip and system levels has become the bottleneck for the improvement in overall system performance and integration [2-4]. The design and fabrication of electronic packages, that can provide the required interconnection wirings, reliable power distribution, cooling, and protection for advanced electronic systems, are becoming increasingly complex and challenging.

In typical electronic packages, around 80% of electronic components are passives, and they can take up to 50% of printed wring board area [5]. The ratio of passive to active components in recent electronic products is rapidly increasing, as shown in Table 1.1.

Table 1.1: Comparison of the number of active (ICs) and passive components

Product	IC	Passive components	Total components	Passive:active
Desktops	182	1066	1285	6:1
Laptops	53	820	900	15:1
PCS Phone	15	322	380	21:1
Pen Pager	15	437	460	29:1

Source: NEMI 2000 Roadmap

Along with the increasing ratio of passive to active components, the total number of passives present in recent electronic products is also steadily increasing. Hence, the technological progress towards developing compact, lighter and reliable electronic systems with improved performance is dictated by: "How well and efficiently can the passive components be integrated in packaging modules of high performance electronic systems" [5].

Among various passive components (capacitors, resistors and inductors), capacitors are the major components in terms of number, size, and weight. Wide ranges of capacitances are needed (from few pF to μF range) for various functions such as power decoupling, switching noise suppression, by-pass filtering, tuning, feedback etc. in electronic circuitry. Progress towards achieving novel compact packaging modules with high packaging densities will be facilitated by the integration of capacitors in thin film form.

In the following section, the current capacitor technology (discrete surface mount capacitors) is compared with the embedded (integrated) thin film capacitor technology. The prime motivations behind using embedded thin film capacitors and the leverages offered by them over discrete surface mount capacitors are discussed in detail.

1.1   Current Capacitor Technology and Trends

(a)   Surface-mount Discrete Capacitors

Around 95% of the capacitors currently used in electronic circuits are discrete capacitors based on surface mount technology (SMT), where a singular component is enclosed in a single case that is mounted on an interconnecting substrate. A photograph of an IBM state-of-the-art multichip module (MCM) is shown in Fig. 1.1(a). The squares are the IC chips (in 11 x 11 array) surrounded by discrete surface mount decoupling capacitors (shown in small dots) [8]. A blown-up view of a discrete surface

2

mount decoupling capacitor is shown in Fig. 1.1(b); these capacitors have millimeter-scale x, y, and z dimensions. These capacitors are connected to the ICs using wire connections. The footprints of the IC and decoupling capacitors are shown in Fig. 1.1(c).

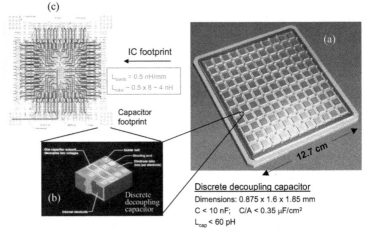

Figure 1.1 **Photographs of IBM's multi-chip module showing the (a) arrangement of IC chips surrounded by the decoupling capacitors, (b) blown-up view of surface mount discrete decoupling capacitor, and (c) IC and capacitor footprints [8].**

(b)   Embedded Capacitors

Embedded thin film capacitor technology, where capacitors can be integrated within the chip or packaging module or printed wiring board (PWB), offers a promising solution to achieve novel compact packaging modules with high packaging density, high performance and reliability along with low system cost [5-8]. Fig. 1.2 shows schematics of two different schemes, where capacitors are integrated a) within the substrate of the packaging module underneath the IC, and b) within the multi-level metallization (MLM) of the IC chip.

3

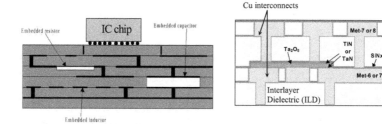

Figure 1.2. **Schematics showing the integration of thin film capacitors (a) within the substrate of the packaging module (off-chip integration) and (b) within the multilevel metallization (MLM) of the IC chip (on-chip integration).**

1.2 Integration of Thin Film Capacitors: Motivations and Leverages

The integration of thin film capacitors is aimed to achieve a) lower cost (per capacitor) associated with the fabrication and placement of capacitors at their specified positions, b) improved performance due to the 2 to 3 orders of magnitude reduction in parasitic inductances of the IC to the capacitor connections, and c) compact electronic packages with higher packaging densities.

(a) Cost Leverages

With SMT discrete capacitors, each capacitor has to be picked up, aligned with the solder balls, and soldered at their respective positions. The handling and positioning of discrete capacitors on individual basis not only involves lot of pick-up and placement costs, but also makes the overall assembly process slow and reduces the total throughput.

In contrast, thin film embedded capacitors can be fabricated using conventional IC processing steps. Using embedded thin film capacitor approach, the individual handling and placements of capacitors can be avoided. Once the process is well developed, several capacitors can be fabricated over the entire wafer using same lithography, developing and etching steps. This approach not only eliminates any individual handling of the capacitors, but also positions all the capacitors in a single step with sub-micron precision. Thus, with capacitor count increasing rapidly in recent electronic products, embedded thin film capacitor technology offers a tremendous cost leverage over currently used surface mount discrete capacitors.

(b)   Easy Integration and Higher Packaging Densities

The surface-mount discrete capacitors (as shown in Fig. 1.1(b)) have millimeter-scale x, y, and z dimensions, which makes their integration within or under the IC chip almost impossible. The large size of discrete capacitors (3-D form factor) has become a limitation in achieving high packaging efficiency, and the cost effectiveness of surface mount discrete capacitors is diminishing, especially when the capacitor count per application-circuit is increasing. In order to meet higher component needs, many miniaturized surface mount discrete capacitors have been mounted on boards and modules in a hybrid fashion. Such surface mounting introduces additional expense of manufacturing, larger board area, and additional parasitics into the system, thus limiting the system performance.

Thin film capacitors have negligible "z" dimension compared to their "x" and "y" dimensions. Hence, higher packaging densities can be achieved by integrating thin film capacitors within the IC multilevel metallization, chip or packaging module or printed wiring board (PWB). Thus, the integration of capacitors offers a promising solution to achieve novel compact packaging modules with higher packaging density and lower volume and weight [9-14].

(c)   Improved Performance and Better Power Decoupling

The rapid developments in very large scale integrated (VLSI) electronics and very high-speed integrated circuits (VHSICs) have brought power supply (distribution) noise regulation issues into the spotlight [10, 15-18]. The power distribution noise (ΔI noise) is generated due to the simultaneous switching of numerous drivers in digital or mixed signal application circuits, which give rise to the voltage fluctuations across the power supply buses. This noise can also couple through quiet logic circuit drivers and appear as spurious voltage signals on the input terminals of logic receiver circuits; such spurious signals may cause erroneous switching of the circuits [18, 19]. The magnitude of these fluctuations increases with the increase in the amount of current drawn during circuit switching, decrease in rise time, and increase in effective chip and package inductances [15]. On the other hand, the improved performance of high-speed systems requires decrease in rise time (in sub-nanosecond range) and increase in IC circuit density. In addition, trends in chip design are such that the voltage and the current signal levels are decreasing according to the CMOS scaling rules, so power supply noise budget must also decrease for the future electronic systems [20].

One effective way to suppress the power distribution noise described above is to provide decoupling capacitors physically close to the switching circuits to supply the current needed during switching [13, 14, 21-23]. Decoupling capacitors reduce power supply level fluctuations by charging up during the steady state and then by supplying the instantaneous current required during simultaneous circuit switching. Thus, they decouple the instantaneous current flow during circuit switching from the board (PWB) and package inductances. In multichip modules (MCMs), surface mount discrete chip capacitors are employed for power decoupling [8, 13]. Typically, surface mount discrete chip capacitors are mounted on the substrate at the corners of each IC chip site. These mounting locations minimize the total inductance of the current supply loop, which consists of the decoupling capacitor, leads connecting the decoupling capacitor to the IC chip containing switching circuits, and ground return paths. However, discrete chip capacitors may not be sufficient for decoupling applications at high operating frequencies because of the large parasitic inductances (~5-10 nH) of the chip-to-capacitor connection leads (as shown in Fig. 1.3). With clock frequencies in GHz range (sub-nanosecond rise time), bus speeds going beyond 100 MHz, and increasing IC gate densities, the inductive parasitics are becoming an important issue.

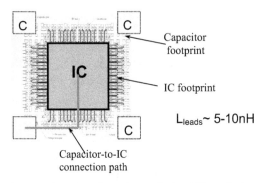

Figure 1.3. **Schematic showing the footprints of the IC and decoupling capacitors along with the capacitor-to-IC connection path [8].**

These limitations can be overcome with the decoupling capacitors in thin film form, integrated into the substrate beneath the chip. By using the embedded thin film capacitor approach, compact packaging modules can be realized, and parasitic inductances can be significantly reduced (by 2 to 3 orders of magnitude) due to shorter connections and elimination of leads. Thin film embedded capacitor technology can be extended to provide on-chip power bus decoupling to reduce ΔI noise due to parasitic inductances of on-chip power distribution wiring at high frequencies, by integrating capacitors into the IC multilevel metallization. This extension of embedded capacitor technology from package to chip level is promising for advanced packaging applications, especially when distinctions between packages and chips are becoming less clear and many package-level problems are appearing at the chip-level with ICs becoming larger and faster.

1.3   Issues and Challenges

Highest possible breakdown field, maximum dielectric constant along with minimum leakage current density, dielectric loss, and dielectric dispersion are the most desirable electrical properties of thin film high-k dielectrics for embedded capacitors. However, there exists a trade-off among the electrical properties of thin film dielectric materials, which determines whether or not a dielectric material can optimally be used for a particular application.

Different applications of embedded capacitors in metal-insulator-metal (MIM) configuration require stringent control on different set of electrical properties. The capacitance of the embedded decoupling capacitors should be sufficiently large such that the energy stored in them can support the current transients caused by many simultaneously switching off-chip drivers and any on-chip switching. Based on the guidelines provided by the Semiconductor Industry Association (SIA), Chahal et $al.$ [21] estimated that capacitance densities in the range of 13-72 nF/cm^2 are required for hand-held and cost performance applications over the next decade. Low series inductance (in pH range) facilitates the reduction in power supply (distribution) noise. Higher breakdown voltage (> 10 V) and lower leakage current density (< 10^{-6} A/cm^2 at 1 MV/cm) are required for both functionality and reliability viewpoints.

Tremendous interest is growing these days in extending the embedded thin film capacitor technology from the realm of well-developed signal electronics domain to the power and automotive electronics applications [24]. Snubbing capacitors integrated either into power devices or into device lead structure have at least two orders of magnitude lower parasitic lead inductances. For most of the power and automotive electronics applications, unlike signal electronics applications, thin film capacitors must comply with an additional requirement of high breakdown voltages, varying from 10 V to 1 kV, depending upon the specific application.

8

Other applications, e.g., blocking capacitors, which separate two dc circuits, require severe limits on leakage current density along with high capacitance and low series inductance. Capacitors used in RF circuits need to have low voltage coefficients (< 100 ppm/V), minimum dispersion at GHz range, good capacitor matching, and precise control of capacitor values. Capacitors integrated in DRAMs require high charge storage density (as high as 25 fF/μm^2) along with extremely low dc leakage (< 10-100 pA at 5V) [5].

Combining all these requirements into one capacitor puts a great deal of demand on the dielectric materials. The development of novel high dielectric constant (high-k) materials that are compatible with already established silicon and polymer (organic substrates) processing technologies is the key driver for the research in capacitor technology. The process compatibility of the deposition techniques with IC and polymer processing facilitates rapid integration by the processing industry because of the reduced risks. One of the major issues with integrating decoupling capacitors in the advanced packaging modules composed of polymeric materials is the requirement of low processing temperatures, which is dictated by the thermal stability of the polymers. Integration of decoupling capacitors in multilevel metallization for on-chip decoupling is also limited by the low thermal budgets for back end of the line processing. Hence, thin film deposition processes with low processing temperatures need to be developed. In addition, process development will be aided by a sound understanding of the effects of various process parameters on the material and electrical properties of the deposited films.

The electrical performance of embedded capacitors also depends on the properties of the overall capacitor structure and their placement/integration schemes. The placement schemes of embedded capacitors in packages or chips should be designed to minimize the parasitic series inductance in the pH range. The extent of interaction (reaction and/or diffusion) of different metal electrodes with dielectric films

9

needs to be evaluated in order to ensure the long-term electrical reliability of the overall capacitor structures. Minimum interface roughness is desired to avoid any premature electrical breakdown of the dielectric films. In addition, the electrode metals have to be selected to avoid poor interlayer adhesion.

1.4   Thesis Focus: Goals and Approach

The overall objective of this research is to explore and quantify the potential of embedded thin film capacitors in terms of their electric performance and to establish key inter-parameter relationships, with experimental emphasis on reactively sputtered tantalum oxide capacitors. In this thesis, reactively sputtered tantalum oxide (Ta_2O_5) thin-films are developed and employed as a cost effective solution for the integration of a high-k dielectric films into established interconnect technologies of IC chips and packages. The electrical performance of these capacitors is also compared with the well established SiO_2 and Si_3N_4 thin film capacitors. These high-k films can be integrated into embedded capacitors operating over a wide range of voltages and in different application geometries (2D and 3D). These films can also be used as dielectric layers for providing effective decoupling of power distribution planes.

1.4.1   Choice of Dielectric Material

This thesis focuses on tantalum oxide (Ta_2O_5) as a dielectric material for embedded thin film capacitors. The prime reason for the choice of this material is its relatively high dielectric constant (~ 25 for amorphous films compared to ~ 3.9 for SiO_2 and ~ 6.9 for Si_3N_4), that is nearly constant through microwave frequencies (up to at least 20 GHz) [25,26], and its compatibility with silicon and polymer flex technologies [27-29]. Although ferroelectric materials such as polycrystalline barium titanate exhibit very high dielectric constants (~ 300 for $BaTiO_3$) at low frequencies, they show dielectric dispersions in GHz frequency range causing a huge drop of dielectric constants from (300 to 18 for BaTiO₃) [30, 31]. Also, deposition of thin ferroelectric films with low leakage current densities at temperatures below 500 °C is difficult, precluding their use on many common microelectronic substrates [32, 33]. Other perovskite materials such as lead lanthanum titanate (PLT) [34], lead lanthanum zirconate titanate [35], and lead zirconate titanate [36] show large variations in dielectric constant with temperature and voltage.

1.4.2 Choice of Deposition Technique

Tantalum oxide films have been deposited using various techniques, such as anodization of thin tantalum films [37, 38], thermal oxidation [39], chemical vapor deposition (CVD) [40-43], radio frequency (RF) or dc reactive sputtering [44-47], and sol-gel processes [48]. Pulsed dc reactive magnetron sputtering [49-51] has been developed and investigated as a way of depositing high quality compound films of desired compositions at low processing temperatures. Pulsed dc reactive magnetron sputtering is similar to dc sputtering except that the polarity of the power supplied to the target oscillates from negative to positive to neutralize the charge build up on the oxide layer at the target. Pulsed dc reactive sputtering is a low temperature deposition technique (< 100°C); as are anodization, sol-gel, and radio frequency (RF) sputtering; these techniques are suitable for applications requiring low thermal budget, such as multichip module (MCM) systems composed of polymeric materials.

While anodization is difficult to integrate with other microelectronic fabrication processes, reactive sputtering is compatible with microelectronic processing techniques. While chemical vapor deposition (CVD) and sol-gel based films tend to have organic and organometallic contaminants from the precursors, sputtering avoids this issue. Unlike radio frequency (RF) sputtering, which suffers from low deposition rates and high capital cost for the impedance matching networks, pulsed dc reactive sputtering can be readily employed to deposit Ta_2O_5 films at relatively high deposition rates and low original system costs.

1.4.3 Approach

The overall electrical performance of thin film capacitors is dictated by the a) intrinsic material and electrical properties of the dielectric film and b) the nature of the metal-dielectric interfaces present in the metal-insulator-metal (MIM) structures of the capacitors (e.g., factors like interface roughness, chemical compatibility of metal to dielectric, etc.).

In this thesis, the research to develop Ta_2O_5 thin film decoupling capacitors is divided into three main thrusts.

(a) Control and optimization of reactive sputtering process to deposit stoichiometric tantalum oxide films free from metallic tantalum or sub-oxides of tantalum. The process variables are optimized to deposit Ta_2O_5 films at highest deposition rates, exhibiting superior and reliable dc breakdown fields and leakage current densities.

(b) Broadband (from dc to 20 GHz) electrical characterization/evaluation of Ta_2O_5 based thin film capacitors. The electrical properties evaluated are breakdown field and leakage current density at dc bias and dielectric constant, dielectric loss, and dielectric dispersion, from dc to 20 GHz.

(c) Evaluation of the effect of properties of capacitor structures (e.g., metal-dielectric interface roughness, metal diffusion, etc.) on their electrical properties.

(a) Optimization of the Deposition Process

A major challenge in the pulsed dc reactive sputtering deposition of tantalum oxide films is selecting the optimum process conditions to deposit stoichiometric films that exhibit superior and reliable electrical properties while maintaining high deposition

12

rates. The stoichiometry of the deposited film, and hence the electrical properties of reactively sputtered tantalum oxide, e.g., breakdown field, leakage current density, dielectric constant, dielectric loss, etc., strongly depend on the oxygen flow rate, sputtering power or magnitude of sputtering ion current, deposition temperature, etc. The presence of many interdependent variables in a nonlinear process like reactive sputtering makes it difficult to optimize the deposition process based solely on experimental observations.

In this thesis, an analytical approach is used to systematically evaluate, predict and control the electrical characteristics of tantalum oxide films as functions of various process variables. An experimentally verified analytical model for pulsed dc reactive sputtering of tantalum oxide films is described and evaluated. The observed influences of important process variables, like oxygen flow rate and sputtering ion current, on the oxygen partial pressure in the chamber, deposition rate, as well as film breakdown and leakage characteristics are predicted using this model.

(b)   Characterization/Evaluation of Electrical Properties of Ta$_2$O$_5$ films

The leverages offered by the Ta$_2$O$_5$ based thin film decoupling capacitors are quantified by evaluating their electrical properties over a wide range of operating voltages and operating frequencies. The electrical properties evaluated are breakdown field, dielectric constant, leakage current density, dielectric dispersion, and dielectric loss. All the above-mentioned electrical properties are also tested for SiO$_2$ and Si$_3$N$_4$, using the same design test vehicles and similar test conditions. Both SiO$_2$ and Si$_3$N$_4$ are standard dielectrics used in microelectronics industry, and hence are used as benchmarks to establish the potential of Ta$_2$O$_5$ films for decoupling capacitors.

Data available in the literature (along with our experimental results) confirm a trade-off between dielectric constant and breakdown field of thin film dielectrics. This empirically determined interdependency sets the "best one can achieve" (BOCA) limits for breakdown fields using a dielectric films of specific thicknesses. The trade-off between breakdown field and dielectric constant is transformed into BOCA performance of capacitors in terms of breakdown voltage and capacitance densities (both are important engineering variables). The breakdown voltages and capacitance densities of Ta_2O_5 capacitors (over a wide range of film thicknesses) are compared with the predicted with BOCA values. The regions and causes of the deviations from the BOCA performance are identified, and experimentally analyzed in detail (chapter 4).

In order to sustain the advantage of higher capacitance density over the whole range of operating frequencies, high-k films should exhibit minimum dielectric dispersion (drop in dielectric constant due to loss of polarizabiltiy at high frequencies). Ta_2O_5 thin films are tested for dielectric dispersion till 20 GHz of operating frequency. High frequency characterization (45 MHz to 20 GHz) is also used to estimate the total parasitic inductance associated with the connection vias present in embedded capacitors (chapter 5).

(c)   Properties of the Capacitor Structure and Key Integration Issues

Thin film decoupling capacitors integrated into an IC's MLM require the compatibility of high-k dielectric with copper, especially when copper is replacing aluminum as a choice of metallization in BEOL (back end of the line). In order to realize the integration of Ta_2O_5 films, either in embedded capacitors or as a dielectric for providing effective decoupling of power distribution planes, its compatibility, reliability and interaction with copper metallization is evaluated. The extent of diffusion of Cu in Ta_2O_5 films and measures to contain this diffusion are experimentally determined. The metal-dielectric interface roughness and microstructure of the dielectric films can dramatically degrade the dc breakdown and leakage characteristics of Ta_2O_5 films. The evolution of the interface roughness with deposition time is theoretically simulated and experimentally verified.

14

1.5   Thesis Overview and Structure

This thesis is organized into 7 chapters. Chapter 2 briefly reviews and compares various thick and thin film capacitor technologies. Various choices of dielectric materials and their deposition processes are also discussed in this chapter.

In Chapter 3, the control and process optimization of reactive sputter process to deposit Ta_2O_5 films with superior and reliable electrical properties; such as breakdown field, leakage current density, and dielectric constant, are discussed. In this chapter, the pulsed dc reactive sputtering process to deposit tantalum oxide films is described using an analytical model. This model is used to predict and analyze the effects of oxygen flow rate and sputtering ion current, on the steady-state oxygen partial pressure, deposition rate as well as the breakdown field and leakage characteristics of the deposited films. The predicted results using an analytical model are also experimentally verified.

In Chapter 4, a design space based on an empirically determined relationship between breakdown field and dielectric constant of film is established and explored. The developed design space predicts the BOCA breakdown voltages (for given capacitance densities) using a dielectric of given thickness. The validity of the developed design space is verified with tantalum oxide thin films. The causes of deviations from BOCA breakdown voltages at different regions of film thicknesses were identified and breakdown voltages were restored to their BOCA values.

In Chapter 5, electrical characterization results of Ta_2O_5 capacitors are presented over a wide range of film thicknesses and operating frequencies (dc to 20 GHz). The electrical properties evaluated in this chapter are dielectric constant, dielectric dispersion, dielectric loss, breakdown field, and leakage current density. These electrical properties are also tested and compared with same design SiO_2 and Si_3N_4 capacitors as benchmarks. Inductances of power connect vias are also determined.

In Chapter 6, the extent of electrode metal diffusion (Cu, Al, Ti and Ta) into Ta_2O_5 is determined using bias-temperature-stress (BTS) and triangular voltage sweep (TVS) techniques. The measures to contain Cu-diffusion into Ta_2O_5 are also presented.

In Chapter 7, the key results and conclusions of this thesis are summarized and discussed. Remaining challenges related to this thesis work are also listed.

1.6 References

1.1   S. A. Campbell, *"The Science and Engineering of Microelectronic Fabrication,"* Oxford University Press, 2001, ch. 9, pp. 205-231.

1.2   R. R. Tummala, *"Fundamentals of Microsystems Packaging,"* McGraw-Hill, 2001, ch. 2, pp. 44-79.

1.3   R. R. Tummala, E.J. Rymaszewski, and A.G. Klopfenstein, *"Microelectronics packaging handbook,"* 2nd ed., Chapman & Hall, New York, 1997, vol 2, ch. 1.

1.4   D. P. Seraphim, R. Lasky, and C. Y. Li, "Principles of Electronic Packaging," McGraw-Hill, 1989, ch. 1, pp. 1-15.

1.5   R. R. Tummala, *"Fundamentals of Microsystems Packaging,"* McGraw-Hill, 2001, ch. 11, pp. 422-464.

1.6   R. C. Frye, Int. J. Electron. Packag. **20** (4), 578 (1997).

1.7   R. R. Tummala, G. E. White, V. Sundaram and S. Bhattacharyam, Advancing Microelectronics **27** (1), 13 (2000).

1.8   J. N. Humenik, J. M. Oberschmidt, L. L. Wu, and S. G. Paull, IBM Journal of Research and Development **36** (5), 935 (1992).

1.9   H. B. Bakoglu, *"Circuits, Interconnections, and Packaging for VLSI,"* Addison-Wesley VLSI Systems Series, ch. 7, pp. 281-335.

1.10   B. K. Gilbert and G.- W. Pan, IEEE Trans. Microwave Theory Tech., **45**, 1819 (1997).

1.11   K. Y. Chen, W. D. Brown, L. W. Schaper, S. S. Ang, and H. A. Naseem, IEEE Trans. Adv. Packag., **23**, 293 (2000).

1.12   R. Dowining, P. Gebler, and G. Katopis, IEEE Trans. Comp., Hybrids, Manufact. Technol., **16**, 484 (1993).

1.13   M. P. Goetz, IEEE Trans. Comp., Packag., Manufact. Technol. B. **19**, 518 (1996).

1.14   P. Jain, J. Y. Kim, Y. Xiao, R. Natarajan, E. J. Rymaszewski, R. J. Gutmann, and T. P. Chow, in *Proc. CPES annual review*, 2000, Blacksburg, VA, pp. 155-158.

1.15   R. R. Tummala, E. J. Rymaszewski, and A.G. Klopfenstein, "*Microelectronics Packaging Handbook*," 2nd ed., vol 2, New York: Chapman and Hall, 1997, ch 3, pp. 232-270.

1.16   R. R. Tummala, "*Fundamentals of Microsystems Packaging*," McGraw-Hill, 2001, ch. 4, pp. 120-182.

1.17   W. D. Brown, "*Advanced Electronic Packaging with Emphasis on Multichip Modules*," IEEE Series on Microelectronic Systems, 1998,ch. 3, pp. 133-136.

1.18   R. K. Poon, "*Computer Circuits Electrical Design*," Prentice Hall, Inc., 1995, ch. 8, pp. 230-271.

1.19   H. H. Chen and J. S. Neely, IEEE Trans. Comp., Packag., Manufact. Technol. B. **21**, 209, (1998).

1.20   International Technology Roadmap for Semiconductors, Assembly and Packaging, 2001 Ed., http://public.itrs.net/files/1999_SIA_Roadmap/Assembly.pdf.

1.21   P. Chahal, R. R. Tummala, G. Allen, and M. Swaminathan, IEEE Trans. Comp., Packag., Manufact. Technol. B. **21**, 184 (1998).

1.22   K. Y. Chen, W. D. Brown, L. W. Schaper, S. S. Ang, and H. A. Naseem, IEEE Trans. Adv. Packag. **23**, 293 (2000).

1.23   R. Senthinathan and J. L. Prince, IEEE Trans. Solid-state Circuits **26**, 1724 (1991).

1.24   D. Van Wyk and F. C. Lee, *Proc. 1999 Annual Power Electronics seminar (CPES)*, Sept.19-26, VA, pp. 62-70.

1.25   J.-Y. Kim, A. Garg, E. J. Rymaszewski, and T.-M. Lu, IEEE Trans. Comp. Packag. Technol. **24** (3), 526 (2001).

1.26   J.-Y. Kim, Ph.D. thesis, Rensselaer Polytech. Inst., Troy, NY, 2000.

1.27   M. C. Nielsen, Ph.D. thesis, Rensselaer Polytech. Inst., Troy, NY, 1998.

1.28   K. Chen, M. Nielsen, S. Soss, E. J. Rymaszewski, T.-M. Lu, and C. Wan, IEEE Trans. Comp., Packag., Manufact. Technol. B. **20**, 293 (2000).

1.29   J.-Y. Kim, M. Nielsen, S. Soss, E. J. Rymaszewski, and T.-M. Lu, J. Appl. Phys. **87** (3), 1448 (2000).

1.30   W.-T. Liu, S. Cochrane, S. X.-M. Wu, P. K. Singh, P. K. Zhang, D. B. Knorr, J. F. McDonald, E. J. Rymaszewski, J. M. Borrego, and T.-M. Lu, Electron. Dev. Lett. **30**(2), 117 (1994).

1.31   P. K. Singh, S. Cochrane, W.-T. Liu, W.-T. K. Chen, D. B. Knorr, J. M. Borrego, E. J. Rymaszewski, and T.-M. Lu, Appl. Phys. Lett.. **66**(26), 3683 (1995).

1.32   R. Ulrich and L. Schaper, *Proc. 2000 International Symp. on Advanced Packaging Materials*, pp. 38-43, 2000.

1.33   G. B. Alers, R. B. van Drover, L. F. Schneemeyer, L. Stirling, C. Y. Sung, P. W. Diodato, R. Liu, Y. H. Wong, R. M. Fleming, D. V. Lang, and J. P. Chang, IEDM- 99, pp. 797-800, 1999.

1.34   S. K. Dey and J.-J. Lee, IEEE Trans. Electron Devices **39**, 1607 (1992).

1.35   D. Dimos, S. J. Lockwood, R. W. Schwartz, and M. S. Rodgers, IEEE Trans. Comp., Packag., Manufact. Technol. A(**18**), 174 (1995).

1.36   W. Williamson III, B. K. Gilbert, H. D. Chen, K. R. Udayakumar, L. E. Cross, and C. M. Bozler, Integr. Ferroelect **10**, 335 (1995).

1.37   P. Wilcox, J. Vac. Sci. Technol. **9**, 74 (1971).

1.38   S. G. Byeon and Y. Tzeng, IEEE Trans. Electron Devices **37**, 972 (1990).

1.39   G. S. Oehrlein, J. Appl. Phys. **59** , 1587 (1986).

1.40   K.-A. Son, A.Y. Mao, Y-M. Sun, B. Y. Kim, F. Liu, A. Kamath, J. M. White, D. A. Roberts, and R. N. Vrtis, Appl. Phys. Lett., **72**, 1187 (1998).

1.41   C. Chaneliere, S. Four, J. L. Autran, R.A. B. Divine, and N. P. Sandler, J. Appl. Phys. **83**, 4823 (1998).

1.42   J. V. Grahn, P. E. Hellberg, and E. Olsson, J. Appl. Phys. **84** , 1632 (1998).

1.43   F.-C. Chiu, J.-J. Wang, J. Y.-M. Lee, and S. C. Wu, J. Appl. Phys. **81** , 6911 (1997).

1.44   S.-D. Cho and K.-W. Paik, Mater. Sci. Eng. B. **67**, 108 (1999).

1.45   S. Roberts, J. Ryan, and L. Nesbit, J. Electrochem. Soc. **133**, 1405 (1986).

1.46   E. Atanassova, T. Dimitrova, and J. Koprinarova, Appl. Surface Sci **84**, 193 (1995).

1.47   P.-H. Chang and H.-Y. Liu, Thin Solid Films **258**, 56 (1995).

1.48   H. Fujikawa, K. Yamashita, and Y. Taga, Materials Chem. And Phys. **35**, 11 (1993).

1.49   X. M. Wu, P. K. Wu, T.-M lu, and E. J. Rymaszewski, Appl. Phys. Lett. **62** (25), 3264 (1993).

1.50   M. C. Nielsen, J.-Y Kim, E. J. Rymaszewski, T-M Lu, A. Kumar, and H. Bakhru, IEEE Trans. Comp., Packag. Manufact. Technol. B. **21** (3), 274 (1998).

1.51   W. D. Sproul, M. E. Graham, M. S. Wong, S. Lopez, D. Li, and R. A. Scholl, J. Vac. Sci. Technol. A **13**, 1188 (1995).

CHAPTER 2

CAPACITOR TECHNOLOGY: PROCESSES AND MATERIALS

In the microelectronics industry, fabrication of capacitors is primarily based on thick and/or thin film technologies. Thick film structures are typically 10 to 50 μm (0.4 to 2.0 mils) in thickness. Thin films normally range in thickness from sub-nanometer to 1-5μm. In the following section, thick and thin film capacitor technologies are briefly reviewed and compared.

2.1 Thick and Thin Film Capacitor Technologies

2.1.1 Thick Film Capacitor Technology

In thick film technology, the ceramic layer (dielectric in the capacitor) between electrodes is applied as paste (or as "wet slurry") by screen-printing and then squeezed. In a stacked capacitor configuration, the ceramic and the metal ink layers are built in successive fashion with the "wet" process, which requires drying before the next layer [1]. The ceramic must be fired (vitrified) after this step at temperatures ranging from 1100 to 1900 °C [2]. Due to high firing temperatures, metals with high melting points, such platinum, palladium, silver, and their alloys dominate the metallization for electrode patterns. Recent process modifications have now allowed the use of nickel (which is less expensive) and copper electrodes as well [1,3].

Low initial investments and the ease of fabrication are the main advantages of using thick films. Thick films are often used in high-power and high-voltage applications because of their high breakdown voltages. Thick film capacitor technology is well established, and is primarily used in analog applications and in some digital and high frequency applications, where capacitor counts are limited, quick turnaround is a must, and economy dictates the capacitor costs. But, the dimensional accuracies achieved using thick film technology is poor by IC fabrication standards. In addition, high temperature ceramic firing steps limit its potential in applications requiring low thermal budgets. Hence, thin film capacitor technologies are employed where higher dimensional accuracies of feature sizes, low processing temperatures, and high-volume production are needed.

2.1.2 Thin Film Capacitor Technology

The dielectric films and electrodes in thin film capacitor technology are formed by several deposition techniques [4,5]; vacuum evaporation, dc and radio-frequency (rf) sputtering, chemical vapor deposition (CVD), sol-gel process, anodization, etc. The choice of the deposition techniques depends on the selection of dielectric materials to be used in thin film capacitors.

In vacuum evaporation, the source materials are heated in vacuum to their sublimation temperatures using electron beam or by resistive heating. Subsequent condensation of material onto the substrate leads to film deposition. Several materials, such as Al, Cu, Ti, TiO_2, Si_3N_4, HfO_2, CeO_2, etc. (both metals and dielectrics), can be deposited by this technique, The advantage of vacuum evaporation is that it can be performed at room or low temperatures, which is important to deposit amorphous films. It's a low-pressure (typically $< 10^{-5}$ torr) deposition technique and thus deposited films can be relatively free from contaminations. The deposition pressures are dictated by the vapor pressure of the evaporated materials, and hence it is not suitable for materials with low vapor pressures [6].

In sputtering, energetic ions are bombarded onto the metal (or metal compound, e.g., oxide, nitride etc.) target. Target species are liberated, when the energy of the bombarding ions is high enough (greater than work functions), and deposit onto the substrate. Using sputtering, the depositions can be done at temperatures less than

22

200 °C. But, films deposited by sputtering with thicknesses less than around 30 nm are prone to pinholes.

In chemical vapor deposition (CVD) the chemical precursors, diluted using an inert gas (hydrogen, helium, etc.) are vaporized and transported into a reaction chamber. These gases flow over the substrate, which is maintained at relatively high temperature. Reactant molecules are adsorbed on the substrate and subsequently undergo a chemical reaction resulting in the deposition of thin film. The inert gas used to dilute the precursors helps to reduce their concentration and avoid spontaneous reaction. The reaction byproducts are carried away from the growing film by the carrier gas. Using plasma during CVD (as in plasma enhanced CVD) can further enhance reaction rates. The main advantage of CVD is that it can be performed over a wide range of pressures from atmospheric to ultra high vacuum. CVD offers good uniformity over large areas, good control over stoichiometry and excellent step coverage. However, for multi-component systems the process can be complex. The manipulation of toxic, explosive or corrosive gases during the processes, organic contaminants from precursors, and the use of comparatively high temperatures (250-1000 °C) are the main drawbacks of CVD.

In sol-gel processes, inorganic or composite organic-inorganic material based films are made at relatively low temperature. Processing consists of hydrolysis of the constituent molecular precursors and subsequent poly-condensation to a glass-like form (gel). Incorporation of organic and inorganic additives during the process of formation of the glassy network at room temperature is the main limitation of this process.

In an aqueous anodization process, the substrate is immersed in an electrolyte solution and oxidation is carried out using a fixed current density and potential. The main advantage of this process is that the oxide can be grown at room temperature and it is homogeneous with reproducible thickness. However, the anodic oxide formed by this method contains impurities from the electrolyte solution, which increases the leakage current and hence degrades the electrical performance.

23

2.2   Materials Options for Dielectrics for Thin Film Capacitors

There are two broad categories of dielectric materials: ferroelectric and paraelectrics, sometimes also called as 'class 1' and 'class 2' dielectrics, respectively [7-9]. At low operating frequencies (< 10 MHz), ferroelectrics, such as polycrystalline barium titanate (BaTiO$_3$) and lead zirconate titanate (PbZr$_x$Ti$_{1-x}$O$_3$) can exhibit up to three orders of magnitude higher dielectric constants compared to paraelectric materials such as silicon oxide (SiO$_2$), aluminum oxide (Al$_2$O$_3$), and tantalum oxide (Ta$_2$O$_5$). Table 2.1 shows the dielectric constants of selected important ferroelectric and paraelectric materials. The values of the dielectric constants for ferroelectrics shown in Table 2.1

Table 2.1.   Dielectric constants of important paraelectric and ferroelectric dielectrics [7,8].

Dielectric	Composition	Dielectric Constant
Paraelectric materials		
Polyimide [7]	Organic	3.6
Silicon Oxide [7]	SiO$_2$	3.9
Silicon Nitride [7]	Si$_3$N$_4$	6.9
Aluminum Oxide [7]	Al$_2$O$_3$	9.1
Aluminum Nitride	AlN	9
Silicon Carbide (3C and 6H) [8]	SiC	9.7
Amorphous Tantalum Oxide [7]	Ta$_2$O$_5$	25
Ferroelectric materials		
Barium Strontium Titanate [7]	BaSrTiO$_3$	up to 1000
Lead Zirconate Titanate [7]	PbZr$_x$Ti$_{1-}$	up to 2000
Barium Titanate [7]	BaTiO$_3$	up to 5000

are their maximum values for completely oriented bulk material single crystal at low frequency and no bias. Their specific values strongly depend on grain size, crystal orientation, frequency of measurement and electrical bias.

Unlike paraelectrics, ferroelectrics have asymmetrical unit cells at room temperature, i.e., the overall center of positive charge for the distribution of cations is slightly separated from the overall center of the negative charge for the distribution of anions. Under an applied electric field, the unit cell dipoles present in ferroelectrics spontaneously orient themselves parallel to the applied field, and hence exhibit very high dielectric constants at room temperature. In contrast, molecules of paraelectric materials exhibit small polarizations due to induced dipoles by the applied electric field. However, ferroelectrics loose their spontaneous polarizations at higher temperatures (e.g. ~120 °C for $BaTiO_3$) and higher signal frequencies (at several hundred MHz), and behave like paraelectrics under those conditions. As an example, polycrystalline perovskite barium titanate exhibits a high dielectric constant (~300) at low frequencies; however, it shows dielectric dispersion and a decrease in dielectric constant (from 300 to around 18) in the GHz frequency range [10,11]. Other perovskite materials such as lead lanthanum titanate (PLT) [12], lead lanthanum zirconate titanate [13], and lead zirconate titanate [14] show large variations in their dielectric constants with temperature and voltage.

In general, ferroelectric materials must achieve a crystalline or polycrystalline state in order to exhibit higher dielectric constants than paraelectrics. This may require anneal temperatures in the range of 500-700 °C. This precludes their use on many common microelectronic substrates [7, 15] such as polymeric substrates, etc. In contrast, most paraelectric materials can also be deposited at low temperatures, and are suitable for applications where the fabrication schemes demand low processing temperatures. In addition, paraelectric materials such as SiO_2, Si_3N_4, Al_2O_3, and amorphous Ta_2O_5 exhibit stable values of dielectric constants with respect to operating frequency, voltage, and temperature [7, 16-21]. General properties of ferroelectric and paraelectric materials are compared in Table 2.2.

Table 2.2.   Comparison of paraelectric and ferroelectric dielectrics

Property	Paraelectrics	Ferroelectrics
ε_r (dielectric constant)	2- 50	from ~15 up to 1000's
ε_r vs. T	little dependence, < 500 ppm/ °C	highly dependent on T due to crystal phase transitions and ion mobility
ε_r vs. frequency	little dependence	decreases significantly above a few GHz
ε_r vs. bias	no dependence	decreases with dc bias
ε_r vs. film thickness	no dependence	highly dependent due to thickness dependent crystal structures
Dielectric fatigue	none	κ can decrease significantly with cycles and time
ε_r vs. film structure	little or no dependence	must be crystalline to exhibit high dielectric cosntant
Cure requirements	none	500- 700 °C (typically in O_2)

As a result, paraelectrics are more suitable for high tolerance decoupling capacitors operating in the GHz frequency range and other high-tolerance applications such as filtering, timing, RF, wireless etc., where predictable and constant capacitance is required. Ferroelectrics are better for high capacitance applications (operating at low frequencies) such as energy storage where high-tolerance and stability is not as important as long as a minimum amount of capacitance is provided. Currently, ferroelectric materials are widely used in DRAM memory cells [7,18]. Although multilayer dielectric structures using different dielectrics are being explored to achieve high capacitance densities, their processing is not cost-effective [17].

Thus, dielectrics like SiO_2, Si_3N_4, and Ta_2O_5 are suitable candidates for embedded capacitor applications operating over wide ranges of frequencies (dc to GHz range) and voltages. In this thesis, some advantages of using Ta_2O_5 as the dielectric for embedded capacitors are established and experimentally demonstrated by comparing its electrical performance with standard dielectric materials like silicon oxide and silicon nitride. The prime advantage of Ta_2O_5 over SiO_2 and Si_3N_4 is its relatively high dielectric constant (~25 for amorphous films compared to ~3.9 for SiO_2 and ~6.9 for Si_3N_4). The dielectric constant of amorphous Ta_2O_5 remains nearly constant through microwave frequencies (up to at least 20 GHz, as discussed in chapter 5) [10]. Fabrication is process compatible with silicon and polymeric flex technologies [11,13,22,23]. The high value of the dielectric constant is especially useful in decoupling capacitors, because the same capacitance can be obtained with a smaller area, required to reduce real estate on IC chips. Ta_2O_5, a tantalum-based material, has relatively little Cu diffusion through it as opposed to SiO_2 (discussed in chapter 6). In addition, Ta_2O_5 can be deposited by several deposition methods. In the following section, various deposition methods for Ta_2O_5 deposition are briefly reviewed.

2.3 Thin Film Deposition Techniques for Tantalum Oxide

This section presents an overview of different techniques used in the literature for depositing Ta_2O_5 thin films. The techniques addressed here are chemical vapor deposition (CVD), anodic and thermal oxidation, sputtering, and atomic layer deposition (ALD).

2.3.1 Chemical Vapor Deposition (CVD)

Several CVD methods have been employed to deposit Ta_2O_5 films: atmospheric pressure CVD (APCVD) [24-26], low pressure CVD (LPCVD) [27-31], plasma-enhanced CVD (PECVD) [32-36], photo-assisted CVD [37]. LPCVD (and PECVD PECVD are the most commonly employed.

A simple tantalum based gaseous precursor, which is easy to store and evaporate to give Ta_2O_5, is not available. Hence, organometallic precursors have to be used to deposit Ta_2O_5 by CVD. These include precursors such as tantalum dichloro-diethoxy-acetylacetonate $TaCl_2(OC_2H_5)_2C_5H_7O_2$ [26], penta-dimethyl-amino tantalum $Ta(N(CH_3)_2)_5$, tantalum penta-methoxide $Ta(OCH_3)_5$ [30], tantalum penta-ethoxide $Ta(OC_2H_5)_5$, etc. [27]. $Ta(OC_2H_5)_5$ is the most commonly used precursor in CVD processes. The precursors are generally associated with oxygen or nitrous oxide as a source of oxygen atoms, diluted in hydrogen or some inert gas. Koyama et al. [38] investigated the volatilities and thermal properties of various Ta precursors and concluded that $Ta(OC_2H_5)_5$ is the best precursor so far because of its outstanding volatility and superior heat resistance.

The dissociation of $Ta(OC_2H_5)_5$ allows the formation of Ta_2O_5, but at the same time generates a large number of species containing carbon, hydrogen, and oxygen such as H_2O, CO, CO_2, CH_4, C_2H_5OH, etc. [39]. These compounds result from the dissociation of the precursor in the deposition chamber and can lead to high levels of hydrocarbon contaminants in the deposited dielectric layers with degraded electrical properties.

Contamination can be reduced by performing the post deposition annealing of the film in presence of O_2. O_2 annealing treatment has been reported to be efficient to react with C and H containing species in the film. Volatile reaction products desorb, resulting in a cleaner film [28, 40]. However, at high temperatures these impurities might react with each other and form some other compound as opposed to an expulsion of these species. This conclusion is based on the observation that even after post deposition annealing at 800 °C, the carbon and oxygen contents of a film were the same [18]. The other option to reduce the hydrocarbon contamination is to use a precursor

28

which does not contain carbon or hydrogen. $TaCl_5$ [41] and TaF_5 [35] have been used for this purpose. Chlorine based or fluorine based impurities were not observed in the Ta_2O_5 films, which indicates the efficiency of these Ta based precursors. An approximate reaction mechanism for the deposition of Ta_2O_5 by CVD using TaF_5 as the precursor may be

$$5H_2 + 2TaF_5 \rightarrow 10HF + 2Ta$$

$$4Ta + 5O_2 \rightarrow 2 Ta_2O_5$$

Hydrogen gets rid of fluorine by forming HF, while the oxygen reacts with the metallic tantalum to form Ta_2O_5.

In summary, excellent step coverage, good uniformity in film thickness and stoichiometry are the main advantages of using CVD. Handling of corrosive precursors, film contamination due to reaction byproducts, and comparatively higher deposition temperature (> 350 °C) are the main disadvantages of CVD for depositing Ta_2O_5 films.

2.3.2 Oxidation Techniques

Tantalum oxide films can be formed by first depositing tantalum onto the substrate by sputtering and then performing oxidation. Two main methods for oxidation are: anodic oxidation (performed in an electrolyte or gas) and thermal oxidation.

(a) Anodic Oxidation

Anodic oxidation can be done using aqueous or non-aqueous solution [42-44] or by using an oxygen plasma [45-47]. In aqueous phase anodization, the substrate is immersed in an electrolyte solution and oxidation is carried out using a fixed current density and potential. The main growth mechanism for oxide is due to the high field ionic conduction in the oxide. The main advantage of this process is that the oxide can be grown at room temperature and it is homogeneous with reproducible thickness.

29

However, the anodic oxide formed by this method can contain impurities from the electrolyte solution, which increases the leakage current and hence degrades the electrical performance.

In non-aqueous (gas) phase anodization, the electrolyte solution is replaced by oxygen plasma. The plasma is created using a parallel plate structure with an ion cathode as the source of negative ions. When the substrate is floating with respect to the plasma, it is called plasma oxidation and when the substrate is positively biased, it is called plasma anodization. High oxidation rates can be achieved using this technique, but the oxides produced contain significant density of defects, which need to be removed by annealing.

(b)   Thermal Oxidation

The substrate (tantalum) is exposed to the oxygen-containing atmosphere at high temperatures (500-800 °C) [47]. In order to obtain stoichiometric films, oxidation is performed at a temperature above 500 °C. Ta_2O_5 films obtained by this method are chemically stable and reproducible. However, this technique cannot be used with the substrates such as polymers, which degrade at high temperatures.

2.3.3   Atomic Layer Deposition (ALD)

The ALD technique involves chemical reactions at the substrate surface carried out in a controlled manner. This allows the uniform distribution of active species on the substrate. The reactants are injected in the reactor in the form of vapor pulses. In between the two reactant pulses, the reactor is flushed. Hence, ALD can be used to get very uniform films. The thickness of the film depends upon the number of reactant cycles used.

Kukli *et al.* [48] deposited Ta_2O_5 films by ALD using $Ta(OC_2H_5)_5$ and H_2 between 150-450 °C with nitrogen as the carrier and purging gas. The films were

30

amorphous and had low hydrocarbon contamination. Low deposition rate is the main drawback of this method.

2.3.4 Sputtering Techniques

Sputtering techniques are based on the cathodic erosion due to bombardment by ionized gas species from a plasma discharge. Deposition of Ta_2O_5 films can be done in two ways: (a) by direct sputtering, where the target to be sputtered is Ta_2O_5 and (b) by reactive sputtering, where Ta target is sputtered in oxygen ambient. Films deposited by sputtering of Ta_2O_5 target tend to be rich in Ta due to preferential sputtering of "Ta" compared to "O" atoms from Ta_2O_5 [49].

In reactive sputtering process, the stoichiometries of the deposited films and deposition rates depend on the competition between target sputtering rate and the reaction rate (to form tantalum oxide from Ta). The sputtering rate is primarily determined by the sputtering ion current/power. The reaction rate depends on the partial pressure of oxygen in the chamber. There exists a critical value of the partial pressure of oxygen in the deposition chamber above that stoichiometric films are obtained and below which films are sub-stoichiometric [50]. The partial pressure of oxygen in the chamber and the sputtering ion current /power should be adjusted such that reaction rate is at least equal to the sputtering rate. The stoichiometry of the film also depends on the temperature of the substrate during deposition with films becoming sub-stoichiometric, if the temperature of the substrate is more than 100 °C [51]. Therefore, control and optimization of process variables such as oxygen flow rate, sputtering ion current, substrate and target temperatures, and chamber pressure to achieve stoichiometric tantalum oxide films is an experimentally involved process.

The presence of pinholes is the main problem with films deposited by sputtering, especially when the films are very thin (< 30 nm). When film deposition starts, first small islands are formed on the substrate, which grow and eventually cover

31

the entire surface. If the film thickness is low then it is likely that the surface is not entirely covered and this leaves pinholes in the film, which degrade its performance. Another cause of pinholes is the impurities present on the substrate or the air trapped in the film during the film growth. However, thicker films are generally free of pinholes.

Depositions using reactive sputtering can be performed at relatively low temperatures (< 200 °C) [17, 52]. This makes it suitable for applications requiring low thermal budgets, such as systems containing polymers. Unlike CVD, ALD, and anodization techniques, reactive sputtering is relatively clean as only pure Ta metal, oxygen, and argon are used in this process. In addition, reactive sputtering is compatible with rest of microelectronic processing. In addition, if the stoichiometries of deposited films are well controlled, reactive sputtering offers the potential to form compositionally graded films and non-equilibrium compounds.

Hence, due to above-mentioned advantages, reactive sputtering is chosen for Ta_2O_5 deposition in this thesis work. The next section of this chapter describes the specific details reactive sputtering system used in this thesis work.

2.4   Reactive Sputtering System Used in This Thesis Work

2.4.1   System Configuration

The sputtering system used in this study is a dc magnetron sputtering system from CVC (Rochester, NY); a schematic is shown in Fig. 2.1. The tantalum metal target has a 99.99% purity level, is 8" in diameter and 0.25" in thickness. The substrate holder is ~5.5 cm above the tantalum target.

A roughing pump is used to evacuate the chamber to 50-100 mtorr. A cryogenic high vacuum pump is then used to evacuate the chamber to a base pressure below 3×10^{-7} torr. The sputtering gas used is a mixture of oxygen and argon in various proportions. Both gases were introduced into the chamber through mass flow controllers. Pressures in the mtorr range were measured using Pirani gauges, while base pressure was measured using ion gauges.

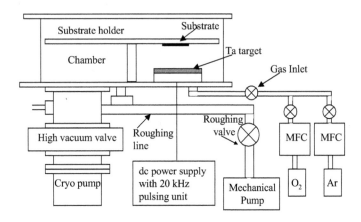

Figure 2.1.   Schematic of the reactive sputtering system used.

2.4.2   Biasing Scheme Used

During reactive sputtering, the Ta metal target reacts with oxygen to form tantalum oxide on the target. This oxide formation on the metallic target is also termed as "target poisoning". The oxide regions of the target do not allow the positive bombarding ions to neutralize their charge and therefore a positive charge layer forms at oxide regions of the target. The effective negative potential of the target thus decreases in magnitude and leads to low deposition rates. The build up of the insulating oxide region is depicted in Fig. 2.2. In addition to low deposition rates, a breakdown can occur between the charges trapped on the top of the insulating oxide film (formed on the target due to target poisoning) and the metallic regions of the target. This breakdown phenomenon creates arcing in the plasma [53-56]. The energy released during arcing can be great enough to eject particulates caused by local melting of the

33

target material. This causes contamination in the deposited films. In terms of electrical properties of the deposited films, a substantial lowering of the breakdown field strength and increase in the leakage current density could arise from the arcing.

A pulsed dc power supply was used in this study to minimize charge build-up and hence the arcing. The periodic pulsing changes the polarity of the bias supplied to the cathode (target). During the negative pulse (large negative voltage), normal sputtering occurs, while during the positive pulse (small positive voltage), the oxide layer on the target is discharged as electrons are attracted to the surface of the oxide.

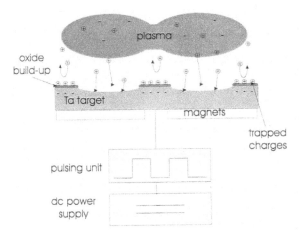

Figure 2.2.   Target poisoning occurs during reactive sputtering with a metallic target. Without the pulsing unit, severe target poisoning will result arcing

An Advanced Energy MDX dc power supply system connected in series with a SPARC-LE arc suppression unit was used to bias the Ta target. The power supply system supplied pulsed power to the target at a 20 kHz frequency. The duty cycle was around 10%, during which the target (cathode) potential shifts to positive voltage ($\sim$ 10 V). The duration of the positive pulse was around 5 μs. The substrate holder and the chamber body were kept electrically grounded. A magnetron system was also used around the target to confine the plasma closer to the target with magnetic fields and achieve higher deposition rates.

2.5   Summary of the Chapter

Thin film capacitor technologies are used in microelectronics applications, where precise dimensional accuracy of features, low processing temperatures, and high-volume production are needed. The choice of deposition technique depends on the choice of dielectric material to be used in thin film capacitors.

In this thesis, Ta_2O_5 is used as the dielectric for thin film capacitors. It has a relatively high dielectric constant ($\sim$ 25), which is constant up to at least 20 GHz. It is also compatible with established silicon and polymer processing technologies. Reactive sputtering is used to deposit Ta_2O_5; this low temperature deposition technique is suitable for applications requiring low thermal budgets.

The next chapter discusses the control and optimization of reactive sputtering process to deposit Ta_2O_5 films with reliable and superior electrical properties.

2.6 References

2.1   C. A. Dostal, Ed., *"Electronics Materials Handbook,"* ASM International vol. 1, 1989, section 2, pp. 255-262.

2.2   P. J. Holmes and R. G. Loabsy, Ed., "Handbook of Thick Film Technology" Electrochemical Publishers, Ltd., 1976, ch.1, pp. 1-20.

2.3   R. R. Tummala, *"Fundamentals of Microsystems Packaging,"* McGraw-Hill, 2001, ch. 11, pp. 420-464.

2.4   S. A. Campbell, *"The Science and Engineering of Microelectronic Fabrication,"* Oxford University Press, 2001, ch. 12 and 13, pp. 283-334.

2.5   S. P. Murarka and M. C. Peckerar, *"Electronics Materials Science and Technology,"* Academic Press, Inc., 1989, ch. 6 and ch. 7, pp. 267-396.

2.6   J. D. Plummer, M. D. Deal, and P. B. Griffin, *"Silicon VLSI Technology Fundamentals, Processing, and Modeling,"* Prentice Hall, NJ, 2000, ch. 9, pp. 530-539.

2.7   R. Ulrich and L. Schaper, *Proc. International Symp. on Advanced Packaging Materials* 2000, pp. 38-43.

2.8   J. F. Shackelford, *"Introduction to Materials Science for Engineers,"* 3rd Ed., Macmillan Publishing Company, New-York, 1992, ch. 11, pp. 555-560.

2.9   D. D. Pollock, "Physical Properties of Materials for Engineers," 2nd Edition, CRC Press, 1993, ch. 12, pp. 554-571.

2.10   W.-T. Liu, S. Cochrane, X.-M. Wu, P. K. Singh, P. K. Zhang, D. B. Knorr, J. F. McDonald, E. J. Rymaszewski, J. M. Borrego, and T.-M. Lu, Electron. Lett. **30**(2), 117 (1994).

2.11   P. K. Singh, S. Cochrane, W.-T. Liu, W.-T. K. Chen, D. B. Knorr, J. M. Borrego, E. J. Rymaszewski, and T.-M. Lu, Appl. Phys. Lett. **66**(26), 3683 (1995).

2.12 S. K. Dey and J.-J. Lee, IEEE Trans. Electron Devices **39**, 1607 (1992).

2.13 D. Dimos, S. J. Lockwood, R. W. Schwartz, and M. S. Rodgers, IEEE Trans. Comp., Packag., Manufact. Technol. A.**18**, 174 (1995).

2.14 W. Williamson III, B. K. Gilbert, H. D. Chen, K. R. Udayakumar, L. E. Cross, and C. M. Bozler, Integr. Ferroelect. **10**, 335 (1995).

2.15 G. B. Alers, R. B. van Drover, L. F. Schneemeyer, L. Stirling, C. Y. Sung, P. W. Diodato, R. Liu, Y. H. Wong, R. M. Fleming, D. V. Lang, and J. P. Chang, *Proc. IEDM- 99*, pp. 797-800, 1999.

2.16 K. Chen, M. Nielsen, S. Soss, E. J. Rymaszewski, T.-M. Lu, and C. Wan, IEEE Trans. Comp., Packag., Manufact. Technol. B **20**, 117 (1997).

2.17 M. C. Nielsen, PhD. Thesis, Rensselaer Polytech. Inst., Troy, NY, 1998.

2.18 C. Chaneliere, J. L. Autran, R. A. B. Devine, and B. Balland, Mater. Sci. Engg. R **22**, 269 (1998).

2.19 S.-D. Cho and K.-W. Paik, Mater. Sci. Engg. **B. 67**, 108 (1999).

2.20 X. M. Wu, P. K. Wu, T. –M. Lu, and E. J. Rymaszewski, Appl. Phys. Lett. **62**(25), 3364 (1993).

2.21 J. Lin, N. Masaaki, A. Tsukune, and M . Yamada, Appl. Phys. Lett. **74**(16), 2370 (1999).

2.22 K. Y. Chen, W. D. Brown, L.W. Schaper, S. S. Ang, and H. A. Naseem, IEEE Trans. Adv. Packag. **23**, 293 (2000).

2.23 J.-Y. Kim, M. Nielsen, E. Rymaszewski, and T.-M. Lu, J. Appl. Phys. **87**(3), 1448 (2000).

2.24 T. Takahashi and H. Itoh, J. Less-Common Met. **38**, 211, (1972).

2.25 W. H. Knausenberger and R. N. Tauber, J. Electrochem. Soc. **120**, 927 (1973).

2.26 E. Kaplan, M. Balog, D. F. Bentchkowsky, J. Electrochem. Soc. **123**, 1570 (1976).

2.27 S. Zaima, T. Furuta, Y. Yasuda, and M. Iida, J. Electrochem. Soc. **137**, 1297 (1990).

2.28 H. Shinriki and M. Nakata, IEEE Trans. Electron Devices ED-38, 455 (1991).

2.29   G. Q. Lo, D. L. Kwong, and S. Lee, Appl. Phys. Lett. **60**, 3286 (1992).

2.30   C. H. An and K. Sugimoto, J. Electrochem. Soc. **139**, 379 (1994).

2.31   J. H. Yun and S. W. Rhee, Thin Solid Films **292**, 324 (1997).

2.32   F. C. Chiu, J. J. Wang, J. Y. Lee, and S. C. Wu, J. Appl. Phys. **81**, 6911 (1997).

2.33   D. Laviale, J. C. Oberlin, and R.A. B. Devine, Appl. Phys. Lett. **65**, 2021 (1994).

2.34   S. K. Jeon, S. W. Han, and J. W. Park, J. Appl. Phys. **77**, 5978 (1995).

2.35   R. A. B. Devine, L. Vallier, J. L. Autran, P. Paillet, and J. L. Leray, Appl. Phys. Lett. **68**, 1775 (1996).

2.36   R. A. B. Devin, Appl. Phys. Lett. **68**, 1924 (1996).

2.37   K. Yamagishi and Y. Tarui, J. Appl. Phys. **25**, 306 (1986).

2.38   H. Koyama, S. Tanimoto, K. Kuroiwa, and Y. Taru, J. Appl. Phys. **33**, 6291 (1994).

2.39   S. B. Desu, Mater. Chem. Phys. **31**, 341 (1992)

2.40   S. Kamiyama, H. Suzuki, H. Watanabe, A. Sakai, H. Kimura, and J. Mizuki, J. Electrochem. Soc. **141**, 1246 (1994).

2.41   H. S. Moon, J. S. Lee, S. W. Han, J. W. Park, J. H. Lee, S. K. Yang, and H. H. Park, J. Mater. Sci. **29**, 3372 (1994).

2.42   J. P. S. Pringle, J. Electrochem. Soc.**119**, 482 (1972).

2.43   J. H. Thomas III, Appl. Phys. Lett. **22**, 406 (1973).

2.44   K. C. Kalra, P. Katyal, J. Appl. Electrochem, **21**, 729 (1991).

2.45   T. A. Jennings, W. Mcniell, R. E. Salomon, J. Electrochem. Soc. **114**, 2134 (1967).

2.46   D. Husted, L. Gruss, T. Makus, J. Electrochem. Soc. **118**, 1989 (1971).

2.47   S. W. Park, H. B. Inn, Thin solid films, **207**, 258 (1992).

2.48   K. Kukli, M. Ritala, M. Leskela, J. Electrochem. Soc. **142**, 1670, (1995).

2.49   L. C. Feldman and J. W. Mayer, "*Fundamentals of Surface and Thin Film Analysis,*" PTR Prentice Hall, 1986, Ch. 4, pp. 69-97.

2.50   P. Jain, V. Bhagwat, E. J. Rymaszewski, T. M. Lu, S. Berg, and T. S. Cale, J. Appl. Phys. **93**(6), in Press (2003).

2.51   P. Jain, J. S. Juneja, V. Bhagwat, E. J. Rymaszewski, and T. M. Lu T, submitted in J. Appl. Phys. (2002).

2.52   J. Y. Kim, PhD. Thesis, Rensselaer Polytech. Inst., Troy, NY, 2000.

2.53   R. J. Shul and S. J. Pearton (Eds.) "Handbook of Advanced Plasma Processing Techniques," Springer, 2000, ch. 2, pp. 33-65.

2.54   R. Scholl, tech. Rep., Advanced Energy Industries, Inc., 1998.

2.55   T. Grove, tech. Rep., Advanced Energy Industries, Inc., 1998.

2.56   S. Schiller, K. Goedicke, V. Kirchhoff, and T. Kpte, *Proc. SVC 38th annual Technical Conf.*, pp. 293-297, 1996.

CHAPTER 3

PROCESS MODELING AND OPTIMIZATION OF REACTIVE SPUTTERING

A major challenge in the pulsed dc reactive sputtering deposition of tantalum oxide films is selecting the optimum process conditions to deposit stoichiometric films that exhibit superior and reliable electrical properties while maintaining high deposition rates. The stoichiometry, and hence the electrical properties of reactively sputtered tantalum oxide, e.g., breakdown field, leakage current density, and dielectric behavior, strongly depend on the oxygen flow rate, sputtering power or magnitude of sputtering ion current, deposition temperature etc.

A large number of experiments would be required to experimentally determine the effect of various process parameters on electrical properties of tantalum oxide films. Since a single parameter cannot be varied independently, optimization of the process based solely on experimental observations is difficult. Deposition of pulsed dc reactively sputtered Ta_2O_5 films with superior and reliable electrical properties requires detailed understanding of the effects of various process parameters on electrical behavior than is currently available. A physics-based approach provides a basis to systematically evaluate, predict and control the electrical characteristics of tantalum oxide films as functions of various process parameters.

In this chapter, first the literature on the various approaches used by different groups to optimize reactive sputtering process parameters for electrically reliable films is briefly reviewed and compared with our approach (section 3.1). An experimentally verified analytical model for pulsed dc reactive sputtering of tantalum oxide films is described in section 3.2. Section 3.3 presents the experimental methodology used for the deposition and electrical characterization of tantalum oxide films. The key results are reported and discussed in detail in section 3.4, where the influences of important

40

process variables, i.e., oxygen flow rate and the sputtering ion current, on the oxygen partial pressure in the chamber, deposition rate, as well as film breakdown and leakage characteristics are predicted using this model. The partial pressure of oxygen during deposition is used to predict the critical oxygen flow rate required to achieve insulating dielectric films (with no metallic tantalum) for a specified sputtering ion current. The experimentally established existence of multiple oxygen partial pressures at a given oxygen flow rate (hysteresis loop) is theoretically explained using steady state analysis. The stabilities of different steady state operating points with respect to fluctuations in oxygen flow rate are discussed. The summary and conclusions of this chapter are presented in section 3.5.

3.1   Historical Background and Our Approach

The optimization of process parameters to deposit tantalum oxide films with high dielectric constants, high breakdown fields and low leakage current densities, has been reported by several groups [1-5]. Schiller *et al.* [1,2] correlated the target voltages with the electrical properties of deposited films, and proposed that the voltage maximum in the current-voltage (I-V) characteristics of the plasma can be used as a guide for the adjustment of the optimum partial pressure of oxygen during deposition. However, Chen *et al.* [3] reported that O_2/Ar ratios, above that recommended by Schiller *et al.* [1,2], should be used to deposit high quality tantalum oxide films. Kim *et al.* [4] proposed the product of dielectric constant and breakdown field as a figure of merit for the electrical quality of tantalum oxide films. They further suggested that the O_2/Ar ratio should be adjusted to maximize this figure of merit. Using a "Design of Experiments" (DoE) approach, Nielsen [5] showed that the percentage of oxygen in the O_2/Ar mixture and the sputtering power (or sputtering current) are the two most important processing parameters affecting the electrical properties of tantalum oxide films.

41

Almost all studies suggest the presence of a critical oxygen percentage at a given sputtering power (or current), below which films exhibit metallic nature resulting in high leakage currents. This regime is called "metallic regime" where target surface is mostly metallic due to insufficient supply of oxygen. Hence, the films deposited in this regime tend to have unoxidized Ta resulting in high leakage currents and very low breakdown fields. But if the oxygen percentage is kept above this critical value for a given sputtering power (or current), deposited films exhibit low leakage currents and high breakdown fields. On the other hand, very high oxygen flow rates give low deposition rates; this regime is referred to "oxide" or "reactive" or "compound regime".

Process development will be aided by a sound understanding of the physical processes responsible for this transition from metallic to oxide regime, and how the position of this transition point varies with system variables. Most studies reported in the literature use experiments to determine the critical oxygen percentages for depositions in the "oxide regime", and the experiments are limited to a specific set of process parameters. Due to the presence of several interdependent process variables, it takes extensive efforts and involved experimentation to optimize the process based solely on experimental observations.

A detailed understanding of the effects of various process parameters on electrical behavior facilitates the deposition of reactively sputtered Ta_2O_5 films with desired electrical properties. A physics-based approach provides a foundation to systematically evaluate, predict and control the electrical characteristics of tantalum oxide films as functions of various process parameters. Berg et $al.$ [6-13] have shown that by considering the total system (target erosion, gas injection, chamber wall deposition, reactive gas gettering on all surfaces, etc.) during deposition, it is possible to develop simple relationships between processing parameters. They showed that process parameters can be tuned to deposit compound films (oxides and nitrides) of titanium, zirconium, aluminum, etc., at stable operating points, and that the stoichiometries of the

42

deposited films can be controlled at a given deposition rate. Similar studies on tantalum oxide films have not been reported in the literature.

In this work, a model for the pulsed dc reactive sputtering of tantalum oxide films based on Berg's analytical description [6] for reactive sputtering is presented. The breakdown and leakage characteristics of tantalum oxide films are correlated with the partial pressure of oxygen in the sputtering chamber during deposition. The results presented in this chapter provide a tool to predict the position and nature of the transition from the "metallic" to the "oxide" regime of sputtering. This, in turn allows to predict the amount of oxygen flow required to reliably deposit tantalum oxide films for a specified sputtering ion current density.

3.2   Reactive Sputtering Model for Ta$_2$O$_5$ Deposition

In this section, a model for pulsed dc reactive sputtering of tantalum in a mixture of argon and oxygen is presented. Oxygen is the reactive gas in this process. A schematic showing the distribution of the oxygen gas flow as it passes through the deposition chamber is drawn in Fig. 3.1.

The incoming molecular oxygen flow (Q_0) is divided primarily into three flows:

(1)   Consumed due to the reaction between tantalum and oxygen at the target, Q_{target}.

(2)   Consumed due to the reaction with tantalum on chamber walls and substrate, which is received as a result of sputtered flux of tantalum from the target, $Q_{chamber}$.

(3)   The flow that is not gettered by any surface of the chamber or target, and passes out of the chamber through the pump, Q_{pump}.

An oxygen mass balance gives

$$Q_0 = Q_{t \arg et} + Q_{chamber} + Q_{pump} \tag{3.1}$$

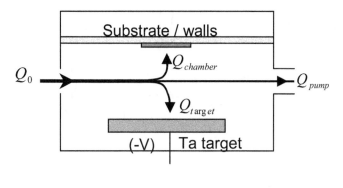

$$Q_0 = Q_{t \, \arg et} + Q_{chamber} + Q_{pump}$$

Figure 3.1. Schematic showing the oxygen flow through the reactive sputtering system. Q_0 is the incoming gas flow of which Q_{target} is consumed at the target, $Q_{chamber}$ is consumed at the chamber walls and substrate and Q_{pump} goes out of the chamber through the pump.

The fractional surface coverage (θ) is defined as the fraction of the surface covered by oxidized tantalum. To simplify the analysis, we assume that stoichiometric oxide (Ta_2O_5) is the only form of oxide that forms as a result of the reaction between tantalum and oxygen. The target has one fractional coverage value, θ_1 and substrate and chamber wall areas have another fractional coverage, θ_2. We also assume that both Ta and Ta_2O_5 fractions on the target are uniformly distributed and well mixed. In our analysis, we ignore the effects of the pulsing of the target bias (at 20 kHz) on the fractional coverages and consider fractional coverages averaged over time scales much larger than the time scale of the pulsing cycle (~50 μs) of the target bias.

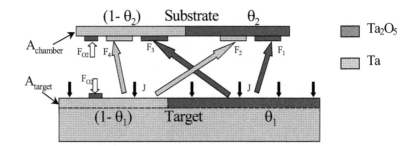

Figure 3.2. Schematic of particle fluxes during reactive sputtering. J is the sputtering ion current density. F_1-F_4 denote fluxes of sputtered material. F_{O2} is the flux of neutral oxygen molecules originating from the oxygen partial pressure p_{O2}. Dotted regions represent the elemental tantalum atoms. Hatched regions represent tantalum oxide molecules.

In Fig. 3.2, J represents the cumulative sputtering ion current density of argon and oxygen ions in the plasma. These positively charged oxygen and argon ions, strike the negatively biased target, giving rise to the flux of sputtered material from the target. The arrows F_1 to F_4, shown in Fig. 3.2, indicate the various types of resultant fluxes of sputtered material onto the substrate and chamber wall areas. Fluxes F_1 and F_4 will not affect the substrate/wall fractional coverage (θ_2). Apart from these four mass fluxes, F_{O_2} denotes the flux of neutral oxygen molecules due to the partial pressure of oxygen gas in the chamber. These molecules move randomly and may strike all surfaces. However, it is assumed that no oxygen is consumed at the fractions of surfaces that are already covered with tantalum oxide (θ_1 and θ_2). If, however, the flux bombards the unreacted elemental Ta fractions ($1-\theta_1$) and ($1-\theta_2$), tantalum oxide (Ta_2O_5) will form as the reaction product. Therefore, F_{O_2} will contribute to an increase in both θ_1 and θ_2.

3.2.1. The Balance of Oxygen Atoms at Target

An expression [6], [14-17] that describes the accumulation of oxygen atoms due to reaction and sputter removal at the target is

$$\frac{dO_{t\,arg\,et}}{dt} = 2F_{O_2}(1-\theta_1)\alpha_{Ta} + 2F_{O_2}\theta_1\alpha_{Ta_2O_5} - 2.5\left(\frac{J}{e}\right)S_{Ta_2O_5}\theta_1 \tag{3.2}$$

where

O_{target} = number of oxygen atoms accumulated with Ta atoms per unit area at the target.

α_{Ta} = sticking coefficient of oxygen molecule to tantalum target.

$\alpha_{Ta_2O_5}$ = sticking coefficient of oxygen molecule to Ta_2O_5.

$S_{Ta_2O_5}$ = average sputtering yield of Ta_2O_5 by incoming argon and oxygen ions.

The factor 2 in the first term of Eq. 3.2 originates from the fact that one oxygen molecule contributes two oxygen atoms to the surface. The factor of 2.5 in the last term of Eq. 3.2 denotes the number of oxygen atoms per tantalum atom in stoichiometric oxide (Ta_2O_5). Under steady state conditions $dO_{target}/dt = 0$ and with the assumption that $\alpha_{Ta} \gg \alpha_{Ta_2O_5}$, the expression for θ_1 can be written as

$$\theta_1 = \frac{2F_{O_2}\alpha_{Ta}}{2F_{O_2}(\alpha_{Ta}) + 2.5\left(\frac{J}{e}\right)S_{Ta_2O_5}} \tag{3.3}$$

3.2.2. The Balance of Oxygen Atoms at Substrate/ Chamber Walls

Similarly, the balance of oxygen atoms at the counter-electrode, which consists of the substrate on which film is to be deposited as well as rest of the chamber wall area where oxygen can deposit, can be performed. The combined area, $A_{chamber}$, is the area on which the material originating from the target as a result of sputtering is deposited. This

46

area is not exposed to energetic ions and hence does not get sputter eroded. The mass balance for oxygen atoms at the counter electrode ($O_{counter\_electrode}$) can be written as

$$\frac{dO_{counter\_electrode}}{dt} = 2F_{O_2}(1-\theta_2)\alpha_{Ta} + 2.5\left(\frac{J}{e}\right)S_{Ta_2O_5}\theta_1\frac{A_{t\arg et}}{A_{chamber}}(1-\theta_2)$$
$$-2.5\left(\frac{J}{e}\right)(1-\theta_1)\,S_{Ta}\frac{A_{t\arg et}}{A_{chamber}}\theta_2 \qquad (3.4)$$

A_{target} is the effective target area for sputtering. The first term on the right hand side of Eq. 3.4 denotes the gain of oxygen atoms due to the reaction of neutral oxygen molecules with the elemental tantalum on the substrate/chamber walls. The second term on right hand side of Eq. 3.4 represents the sputtered flux of Ta_2O_5 (F_3) depositing on the tantalum fraction $(1-\theta_2)$ of the substrate/chamber walls. The third term represents the sputtered flux of tantalum (F_2) deposited on the oxide fraction (θ_2) of the substrate/chamber walls. Under steady state conditions $dO_{counter\_electrode}/dt = 0$, and the fractional coverage θ_2 can be expressed as

$$\theta_2 = \left[1+\frac{2.5\left(\frac{J}{e}\right)(1-\theta_1)\,S_{Ta}\dfrac{A_{t\arg et}}{A_{chamber}}}{2\,F_{O_2}\alpha_{Ta}+2.5\left(\frac{J}{e}\right)S_{Ta_2O_5}\theta_1\dfrac{A_{t\arg et}}{A_{chamber}}}\right]^{-1} \qquad (3.5)$$

3.2.3. The Overall Balance of Oxygen Atoms in the Chamber

From an overall oxygen balance, we get

$$Q_0 = Q_{t\arg et} + Q_{chamber} + Q_{pump}\text{ , where} \qquad (3.6)$$

$$Q_{t\arg et} = \alpha_{Ta}F_{O_2}(1-\theta_1)A_{t\arg et} \qquad (3.6a)$$

$$Q_{chamber} = \alpha_{Ta}F_{O_2}(1-\theta_2)A_{chamber} \qquad (3.6b)$$

47

$$Q_{pump} = \frac{p_{O_2}(S_{pump})}{RT} N_A \qquad (3.6c)$$

where p_{O_2} is the partial pressure of oxygen in the chamber, S_{pump} is the external pumping speed of the pump in m^3/s, N_A represents Avogadro's number, R represents the universal gas constant, (8.314 J/Kgmole), and T is the gas temperature (in K). From kinetic theory of gases, the flux of neutral oxygen molecules and its partial pressure in the chamber are related by

$$F_{O_2} = \frac{p_{O_2}}{\sqrt{(2\pi k T M_{O_2})}} \qquad (3.7)$$

where M_{O_2} is the molecular weight of oxygen, and k represents Boltzmann's constant.

Hence for a given oxygen partial pressure (p_{O_2}) and chamber temperature (T), F_{O_2} can be calculated using Eq. 3.7. Then θ_1 and θ_2 can be estimated from Eq. 3.3 and Eq. 3.5, respectively. Now using F_{O_2}, θ_1 and θ_2, the total flow rate (Q_0) required to achieve this partial pressure, p_{O_2} can be calculated using Eqs. 3.6, 3.6(a), 3.6(b) and 3.6(c).

3.2.4. Deposition Rate

Total erosion rate to the substrate, E, can be calculated using

$$E = \left(\frac{J}{e}\right) \left(\frac{\theta_1 S_{Ta_2O_5} + (1-\theta_1)S_{Ta}}{N_{avg,molar}}\right) \frac{A_{target}}{A_{chamber}} \qquad (3.8)$$

where $N_{avg,molar}$ is the average molecular density (in atoms/cm^3) given by

$$N_{avg,molar} = 6.023 \times 10^{23} \left(\frac{\rho_{Ta_2O_5}}{M_{Ta_2O_5}} \theta_1 + \frac{\rho_{Ta}}{M_{Ta}} (1-\theta_1)\right) \qquad (3.9)$$

where ρ and M are density and molecular weight. We assume that the total erosion rate (E) closely mirrors the total deposition rate to the substrate (D).

48

3.3   Experimental Methodology

3.3.1   Partial Pressure Measurements

The reaction rate to form Ta_2O_5 from Ta depends mainly on the partial pressure of O_2 in the chamber. Hence, controlling O_2 partial pressure in the chamber can control reaction rate.

The following sets of experiments were performed to determine the oxygen partial pressure as a function of oxygen flow rate to the chamber. First, a flow of argon at 17.0 sccm was established such that the chamber pressure was 2.4 mtorr. The throttle valve position was kept at a fixed position. After the pressure reading due to argon attained a steady state at 2.4 mtorr, oxygen gas was introduced into the chamber at controlled flow rates. The steady state total pressure was monitored for each flow rate of oxygen, which varied from 0.0 to 10.0 sccm. The argon partial pressure (2.4 mtorr) was subtracted from the total pressure reading to get the partial pressure of oxygen in the sputtering chamber as a function of oxygen flow rate (Q_0). The additive nature of oxygen and argon partial pressure to give total chamber pressure was verified prior to above-mentioned experiment.

These experiments were performed at constant ion currents and repeated for five different sputtering ion currents, 0.0, 2.0, 3.0, 5.0 and 6.0 A. The substrate was kept electrically grounded during the deposition. Same power supply was used as described in chapter 2 (section 2.4.2). The power supply system supplied a pulsed power to the target at a 20 kHz frequency. The duty cycle was around 10%, during which the target (cathode) potential shifts to positive voltage (~ 10 V). The duration of the positive pulse was around 5 μs. A schematic of the pulsed dc reactive sputtering system is shown in Fig. 3.3.

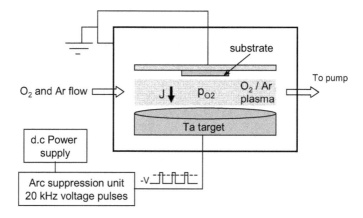

Figure 3.3.   A schematic of pulsed dc reactive sputtering reactor used in this study

3.3.2   Current-voltage (I-V) Characteristics

Metal-Oxide-Metal (MOM) structures were used for the electrical characterization of deposited films. Aluminum, with a target thickness of 0.3 μm, was deposited across the entire wafer, using electron-beam (e-beam) evaporation, to serve as the bottom electrode. 0.3 μm Al provides continuous coverage of bottom electrodes that are free from pinholes and/or hillocks. Different samples of tantalum oxide were deposited at different oxygen flow rates for 2.0 and 5.0 A sputtering ion currents. Sputtering ion currents lower than 2.0 A give very low deposition rates, while higher than 5.0 A lie very close to the maximum limit of the power supply used. The argon pressure was kept constant at 2.4 mtorr in all depositions as low pressures give low deposition rates, while films deposited at higher pressures are prone to contamination. The top aluminum electrodes, with a target diameter of 0.5 mm and target thickness of 0.3 μm, were deposited by e-beam evaporation using a shadow mask. Tantalum oxide films were also deposited on small bare silicon pieces (loaded in chamber along with the metallized sample) at the same time to monitor the film thickness. Film thickness was measured using variable-angle variable-wavelength ellipsometer [18] and also verified by step-height measurements using alpha stepper [19]. Film thicknesses were measured after the films were removed from the deposition chamber.

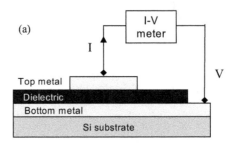

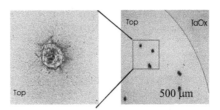

Figure 3.4.   (a) Biasing scheme used to measure dc leakage current density and breakdown field. (b) Top view of capacitor after electrical breakdown.

The current-voltage (I-V) characteristics of the tantalum oxide films were measured on the MOM structure using an HP 4140B I-V meter [20]. Standard pin probes were used to make electrical connections to both top and bottom electrodes of the capacitor structure. The schematic of the electrical testing setup is shown in Fig. 3.4. Voltage was ramped up in steps, with a step value of 1 V, until breakdown of the film occurred. A step delay of 0.5 second was maintained at each voltage step. Electrical breakdown of the film was manifested by the sharp increase in leakage current from around 10^{-6} to 10^{-2} A with small (1 to 3 voltage ramp steps) increase in applied voltage. Films were permanently damaged after electrical breakdown (shown in Fig. 3.4 (b)). The capacitance measurements were done using HP 4280 C-V meter [21] and HP 4192 impedance meter [22] at 1 MHz with 30 mV rms signal level.

3.4   Results and Discussion

3.4.1   Determination of Steady-State Partial Pressure of Oxygen

A steady-state partial pressure of oxygen will be achieved in the deposition chamber once the supply of oxygen balances its total consumption, i.e.

$$Q_0 = Q_{t\arg et} + Q_{chamber} + Q_{pump} \qquad (3.6)$$

The left hand side of Eq. 3.6 (Q_0) represents the supply function (S) of oxygen into the chamber. The flow rate of oxygen, Q_0, can be set using a mass flow controller and hence $S = Q_0$, is independent of chamber partial pressure. The right hand side of Eq. 3.6 represents the consumption function (C) of oxygen, which includes (i) reaction with the target, Q_{target}, (ii) gettering on the chamber and substrate walls, $Q_{chamber}$, and (iii) the oxygen that goes out of the chamber through the pump, Q_{pump}. Unlike the supply function, S, all three parts of the consumption function, C, depend on the partial pressure of oxygen, p_{O_2} and sputtering current density, as given by Eqs. 3.6(a), 3.6(b) and 3.6(c), and 3.7.

For a given sputtering ion current density, the solution of Eq. 3.6 (S = C) will determine the steady-state oxygen partial pressure inside the chamber. Representative curves of supply and consumption functions are plotted in Fig. 3.5 as functions of the oxygen partial pressure, p_{O_2} for a fixed sputtering ion current. For a given oxygen flow rate (Q_0), the supply function S is Q_0 and hence is a horizontal line on Fig. 3.5. The

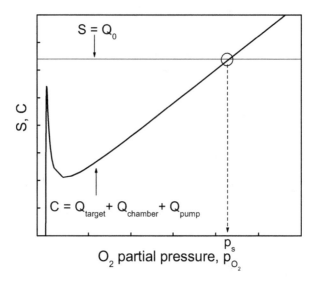

Figure 3.5. Steady-state oxygen partial pressure, p_s at a given oxygen flow rate, Q_0 from the intersection of oxygen supply curve, "S" and oxygen consumption curve, "C".

point of intersection between S and C curves determines the oxygen partial pressure at which the reactor is at steady state. From this point of intersection, steady-state oxygen partial pressure (p_s) can be found by following a vertical line to the p_{O_2} axis as shown in Fig. 3.5.

3.4.2 Multiple Steady States

Suppose we begin to feed the deposition chamber at some relatively low oxygen flow rate, Q_{01}. If the curves for the functions S and C are constructed, as illustrated by curves "Q_{01}" and "y", respectively in Fig. 3.6(a), we see that there is only one point of intersection, point "1". Now, if oxygen flow rate is increased to Q_{02}, the C(p_{O_2}) remains unchanged, but curve S moves vertically up, as shown in Fig. 3.6(a) to line "Q_{02}". Line "Q_{02}" intersects $C(p_{O_2})$ at point "2" and is tangent at point "3". Consequently, there are two steady-state oxygen partial pressures, p_{s2} and p_{s3}, that can be realized at Q_{02}. If the oxygen flow rate is further increased to Q_{03}, the S curve,

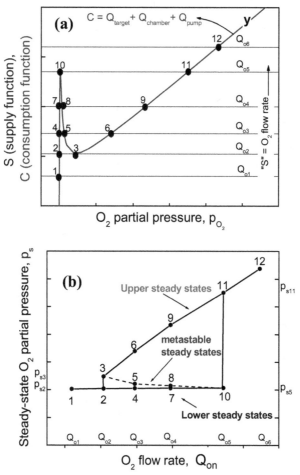

Figure 3.6. Determination of multiple steady states: (a) with the variation of oxygen flow rate,Q_0 and (b) steady-state oxygen partial pressure(p_s) as a function of oxygen flow rate (Q_0), showing upper, lower and unstable steady states.

line "Q_{03}" in Fig. 3.6(a), intersects C(p_{O_2}) three times. Hence there are three different steady-state partial pressures corresponding to Q_{03}. As we continue to increase Q_0, we finally reach line "Q_{05}", at which there are only two steady-state oxygen partial pressures. With further increases in Q_0 we reach line "Q_{06}", corresponding to Q_{06}, where there is only one intersection point. Hence, there exist a range of oxygen flow rates (Q_{02} to Q_{05}) where multiple steady states are manifested. Steady-state oxygen partial pressures corresponding to the six different oxygen flow rates discussed above are summarized in Table. 3.1.

Table 3.1. Steady-state oxygen partial pressures corresponding to the different oxygen flow rates. Multiple steady states are realized between Q_{02} and Q_{05}.

Oxygen flow rate	Steady-state oxygen partial pressure		
Q_{01}		p_{s1}	
Q_{02}	p_{s2}		p_{s3}
Q_{03}	p_{s4}	p_{s5}	p_{s6}
Q_{04}	p_{s7}	p_{s8}	p_{s9}
Q_{05}	p_{s10}	p_{s11}	
Q_{06}		p_{s12}	

By plotting p_s as a function of Q_0, Fig. 3.6(b) can be obtained. From Fig. 3.6(b), as the oxygen flow rate is increased, the steady state oxygen partial pressure increases along the bottom line until Q_{05} is reached. A very small increase in Q_0 beyond Q_{05} causes a sharp rise in oxygen partial pressure from p_{s5} to p_{s11}, as shown in Fig. 3.6(b). On the other hand, if the reactor is operating at p_{s12} and then the oxygen flow rate is decreased down from Q_{06} to Q_{02}, the steady-state oxygen partial pressure p_{s3} is reached. Any slight decrease below Q_{02} and the steady-state oxygen partial pressure drops to p_{s2}. The oxygen flow rates at which sharp changes in oxygen partial pressure are exhibited are called "critical" oxygen flow rates. In a reactive sputtering system, the typical partial pressure versus oxygen flow rate curve shows one "lower" and one "upper" critical oxygen flow rate (Q_{02} and Q_{05}, respectively, in the above discussion). Between the lower and upper critical oxygen flow rates, the reactive sputtering system exhibits multiple steady states in oxygen partial pressure. Experimentally, the hysteresis in oxygen partial pressure versus flow rate curve can be observed depending upon whether the oxygen flow rate is varied in increasing flow rate mode or decreasing flow rate mode. The width of this hysteresis region is determined by the position of lower and upper critical oxygen flow rates.

Oxygen entering the chamber gets consumed in oxidizing either the tantalum target or unreacted tantalum on the chamber walls or substrate to form tantalum oxide. At the same time, this deposited tantalum oxide on the target also gets sputtered off due to the bombardment of the negatively biased target surface by high energy positively charged plasma ions. Hence, there is a competition between the formation of oxide on the target and its removal due to sputtering. A net steady-state fractional coverage at the target (θ_1) and in turn at the chamber walls/substrate (θ_2) is attained, when the sputtering rate balances with the reaction rate.

At low oxygen flow rates, the steady-state fractional coverage at the target (θ_1) is small and hence, most of the target is still metallic Ta. Thus, the majority of the sputtered flux from the target and deposited on the chamber walls/substrate is metallic Ta. Hence, almost all of the incoming oxygen is consumed in oxidizing Ta to tantalum oxide on the target and chamber walls/substrate. Thus, a small amount of O_2 is left to contribute towards oxygen partial pressure. Consequently, low partial pressures of oxygen are exhibited at low oxygen flow rates. At a fixed sputtering ion current, the fractional coverage at the target (θ_1) increases with the increases in oxygen flow rate and thus, more and more fraction of the target is now covered with tantalum oxide. As a result, the amount of Ta that sputters off the target and deposits on the chamber walls/substrate also decreases. Consequently, now less oxygen is consumed in oxidizing metallic Ta to tantalum oxide on the chamber walls/substrate and steady-state oxygen partial pressure starts to increases with oxygen flow rates. This increase in partial

56

pressure causes a further increase in the formation of tantalum oxide on the target. Thus even less oxygen is used and the oxygen partial pressure increases further, etc. The flow rate at which the fractional coverage at the target (θ_1) approaches unity, there is a nonlinear increase in the oxygen partial pressure and a runaway situation forms that shifts the operating point from "10" to "11" in Fig. 3.6(b). An analogous situation happens if the oxygen flow is decreased. Reaching the point "3" (coming from higher values of oxygen flow) causes a similar runaway situation that shifts the operating point from "3" to "2" in Fig. 3.6(b).

3.4.3   Stability of Steady States

The middle points "5" and "8" in Fig. 3.6(a) and Fig. 3.6(b) represent metastable operating points. If by some means the system is operated at p_{s5} and a small change in partial pressure suddenly occurs due to system fluctuations, causing the partial pressure to move slightly towards the right side. At this new condition, the consumption rate C is lower than supply rate S, and that causes further increase in oxygen partial pressure. Oxygen partial pressure continues to increase until partial pressure p_{s6} is reached. On the other hand, if the partial pressure decreases slightly from the operating point p_{s5}, the supply rate (S) is smaller than consumption rate (C) and steady state oxygen pressure continues to fall until p_{s4} is reached. Hence, intermediate steady state conditions in the region where multiple steady states exist are metastable. Operation at these unstable steady states, i.e., the region in which the slope of the partial pressure versus oxygen flow rate curve is negative, is very diffucult. However, with proper sophisticated process control, Hmiel [23] has experimentally demonstrated operation in negative slope portion of pressure versus gas flow curve, for the dc magnetron reactive sputtering of TiN.

Similar stability analyses can be carried out for the other steady states (p_{s1}, p_{s2}, p_{s4}, p_{s6}, p_{s7}, p_{s9}, p_{s11} and p_{s12}). When these partial pressures are subjected to either a small positive and negative fluctuation, they return to their steady state values. Consequently we call these operating points as "stable" steady states.

3.4.4 Sputtering Ion Current

The partial pressures of oxygen as functions of oxygen flow rate for different sputtering ion current are plotted in Fig. 3.7(a)-3.7(d). Oxygen flow rate was first increased from 0.0 to 10.0 sccm (increasing flow rate mode) and then decreased from 10.0 to 0.0 sccm (decreasing flow rate mode). In the absence of any sputtering ion current ($I = 0$ A), partial pressure of oxygen increases linearly with oxygen flow rate and decreases linearly with decreasing flow rate. The slope of these straight lines for $I = 0$ case at both increasing and decreasing flow rate modes is determined by the external pumping speed of the vacuum pump. Four cases with sputtering ion currents of 2.0, 3.0, 5.0 and 6.0 A are shown in Fig. 3.7(a)-3.7(d). The parameters used in the calculations are tabulated in Table. 3.2. Typical values of oxide and metal sputtering yields are used [5]. A_{target}, $A_{chamber}$, and S_{pump} are taken from our reactor system and pump dimensions.

Table 3.2. The values of the parameters used in the proposed analytical model [5]

Parameters	Symbols	Value used
Sticking coefficients of O_2 to Ta	α_{Ta}	1
Sputtering yield of Ta with Ar/O_2 sputtering ions	S_{Ta}	0.6
Sputtering yield of Ta_2O_5 with Ar/O_2 sputtering ions	$S_{Ta_2O_5}$	0.024
Target Area (effective area where sputtering occurs)	A_{target}	100 cm^2
Ratio of target to chamber/substrate area	$A_{chamber}/A_{target}$	3.0
Pumping speed	S_{pump}	8.6 L/s

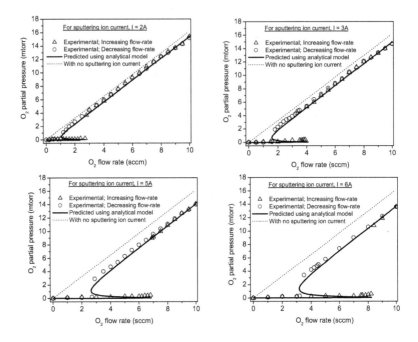

Figure 3.7. Variation of steady-state oxygen partial pressure (p_s) with oxygen flow rate (Q_0) for sputtering ion currents of: (a) 2 A, (b) 3 A, (c) 5 A, (d) 6 A. Solid lines (——) show the predicted curves using analytical model, (Δ) and (O) show the experimentally determined values for increasing and decreasing flow rate modes respectively. Dotted lines (...) are the experimental curves for I = 0 A (no sputtering) case.

The difference between the partial pressure curves at any particular sputtering current and for "no sputtering" case gives an estimate of the amount of oxygen used in oxidizing tantalum to tantalum oxide.

Experimentally determined partial pressure versus flow rate curves compare well with the curves predicted using the analytical model, both in terms of the width of hysteresis loop and the position of the critical oxygen flow rates. The sputtering rate goes up with increasing sputtering ion current. Therefore, higher oxygen partial pressure is needed to balance the reaction rate with the sputtering rate at higher sputtering current. This shifts the critical point (oxygen flow rate at which partial pressure abruptly goes up) towards the higher oxygen flow rate. The width of the hysteresis also increases with sputtering ion current.

3.4.5   Deposition Rate

The total average deposition rates, D (assumed to be equal to the erosion rate, E) as calculated from Eq. 3.8 (normalized wrt. $\theta_1=0$), and the experimentally determined deposition rates for I = 2 A are plotted as functions of oxygen flow rate in Fig. 3.8.

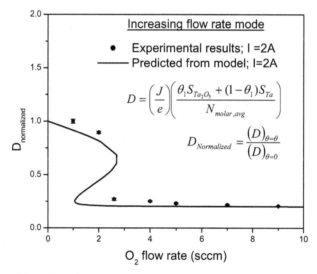

Figure 3.8.   Experimental and predicted normalized deposition rates for I= 2A.

Experimentally determined deposition rates are in good agreement with deposition rates computed using the analytical model. At low oxygen flow rates, the target is partially oxidized and hence most of the sputtered material from the target is metallic tantalum. Typical sputtering yields of metallic tantalum by Ar or O_2 ions are an order of magnitude higher than those for tantalum oxide. The high sputtering yield of tantalum atoms compared to tantalum oxide is responsible for the higher deposition rates at low oxygen flow rates (metallic regime). But once the target gets more fully oxidized, at relatively high oxygen flow rates (oxide regime), the deposition rate comes down. The hysteresis in deposition rate also exists between increasing and decreasing flow rate modes.

3.4.6 Electrical Properties

In this section, we discuss variations in the fundamental dc electrical properties: breakdown field and leakage current density of tantalum oxide films deposited at various process conditions in terms of oxygen flow and sputtering ion currents.

The breakdown fields of the films deposited at different oxygen flow rates (and hence at different partial pressures) are plotted against the oxygen flow rate in Fig. 3.9, for two different sputtering ion currents, 2 A and 5 A. The films were deposited in increasing oxygen flow rate mode; i.e., oxygen flow rate was increased from 0.0 sccm to the required flow rate (in 0.1 sccm steps) and then allowed to stabilize. Argon partial pressure in all the runs was fixed at 2.4 mtorr. For a fixed sputtering ion current, the films deposited at low oxygen flow rates were electrically short (conductive) and showed leakage current density above the current compliance limit of the I-V meter. But after a critical oxygen flow rate, which matches with the predicted upper critical oxygen flow from the partial pressure versus flow rate curve, the deposited films showed excellent breakdown and leakage characteristics. Hence, in order to deposit electrically insulating films, it is important to keep the oxygen flow rate above the upper critical flow rate as predicted using partial pressure versus flow rate curve (as determined from Fig. 3.7).

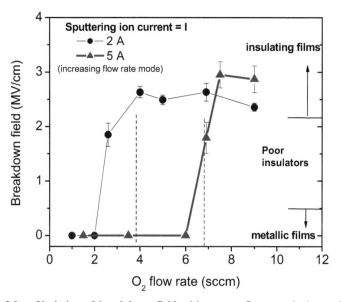

Figure 3.9. Variation of breakdown field with oxygen flow rate in increasing flow rate mode, for I= 2, 5 A. Dotted lines shows the predicted critical oxygen flow rate using analytical model.

The electrical behavior of the tantalum oxide films can be related to the partial pressure behavior as a function of oxygen flow rate. At low oxygen flow rates, the partial pressure of oxygen is low. This indicates that most of the oxygen is being used up in oxidizing tantalum to tantalum oxide, both at the target as well as at chamber walls/substrate. Since the target is only partially oxidized (i.e. in metallic regime), the sputtered flux contains some metallic tantalum and metallic (tantalum rich) films showing nearly zero breakdown field strengths and high leakage current densities are deposited. As oxygen flow rate is further increased at a fixed sputtering ion current, the fractional coverage at the target (θ_1) by oxygen and at the chamber walls/substrate (θ_2) approach unity and hence system approaches towards the oxide regime. This leads to the deposition of stoichiometric tantalum oxide films that show excellent breakdown field strengths and leakage characteristics. The leakage current density decreases from 10^{-2} to 10^{-8} A/cm^2 range at 0.5 MV/cm, as shown in Fig. 3.10. The dielectric constant at 1 MHz also decreases to 24, consistent with the values reported in literature for stoichiometric tantalum oxide films [24]. At a given sputtering current, the position of the critical oxygen flow rate required for the deposition of stoichiometric oxides is nearly the same as the upper critical oxygen flow rate predicted from the partial pressure versus oxygen flow rate curve. Thus, the partial pressure of oxygen during deposition can be used to predict the oxygen flow required to achieve good quality dielectric tantalum oxide films at any sputtering ion current.

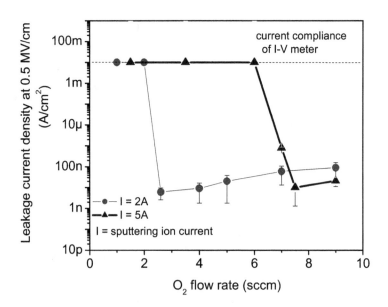

Figure 3.10. Leakage current density of deposited Ta$_2$O$_5$ films at 0.5 MV/cm as a function of oxygen flow rate, for sputtering ion current of 2 and 5 A.

3.4.7 Hysteresis in Breakdown Characteristics

The breakdown fields of the films deposited at a sputtering ion current of 5 A are plotted in Fig. 3.11, for both increasing and decreasing flow rate modes. It is evident that the hysteresis in breakdown field follows the hysteresis in oxygen partial pressure curves. In the hysteresis region, the breakdown fields of the films deposited in decreasing flow rate modes are significantly higher than those deposited in increasing flow rate modes. In the hysteresis region, the films deposited in increasing flow rate mode were found to be electrically conductive while films deposited in decreasing flow rate mode showed relatively good breakdown fields. This hysteresis effect in breakdown fields of the films deposited in hysteresis regime may affect the electrical

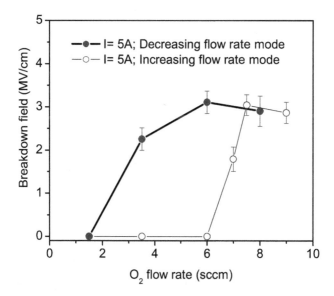

Figure 3.11. Breakdown field as a function of oxygen flow rate for sputtering ion current of 5 A. Breakdown field shows hysteresis effect with respect to oxygen flow rate during increasing and decreasing flow rate modes. The position and the width of hysteresis in breakdown field is same as in partial pressure curves.

reliability, especially if there are fluctuations in oxygen flow rate control system. Hence, based on our setup and experimental results, tantalum oxide films should be deposited at an oxygen flow rate higher than critical oxygen flow rate determined for increasing flow rate mode. At a given oxygen flow rate, higher partial pressure at an upper steady state ensures the higher reaction rate with tantalum to form tantalum oxide, as compared to lower steady state.

3.5   Summary and Conclusions

In this chapter, the pulsed dc reactive sputtering process to deposit tantalum oxide films is described using an analytical model. This model predicts and analyzes the effects of oxygen flow rate and sputtering ion current on the steady-state partial pressure, deposition rate as well as the breakdown field and leakage characteristics of the deposited films. Experimental results are in close agreement to the model predictions.

During a reactive sputtering process, the steady-state fractional coverages at the target (θ_1) and chamber walls/substrate (θ_2) at a fixed sputtering ion current are determined by the net balance of reaction and sputtering rate. The steady-state fractional coverages at the target (θ_1) and chamber walls/substrate (θ_2) then in turn determine the net deposition rate, film composition and hence the breakdown and leakage characteristics of the deposited films. At very low and high oxygen flow rates, single steady-state oxygen partial pressures exist. At intermediate oxygen flow rates, multiple steady states exist at the same oxygen flow rate, *viz.* lower, intermediate and upper steady states. At an upper steady state, a higher oxygen partial pressure is exhibited, while at a lower steady state, a very low oxygen partial pressure is exhibited.

At lower steady states, due to the low oxygen partial pressures, the target is partially oxidized. Thus, the sputtered flux consists of both Ta and TaO_x fragments. The sputtering yield of Ta is an order of magnitude higher than that of oxidized Ta. This causes net higher deposition rate at lower steady states. The presence of metallic tantalum in the films deposited at lower steady states makes them electrically conductive. On the other hand, at upper steady states the target is fully oxidized to tantalum oxide and thus lower deposition rate is exhibited and the deposited films show excellent breakdown and leakage characteristics.

The operating points at upper or lower steady states are stable with respect to any system fluctuations. However, intermediate steady states are unstable to system fluctuations and would toggle to either upper or lower steady state depending upon the nature of the system fluctuation. Hysteresis in partial pressure, deposition rate, breakdown and leakage characteristics during increasing and decreasing flow rate modes is due to the fact that the system follows the lower steady states during increasing flow rate mode and upper steady states during decreasing flow rate mode. The experimental results on our sputtering system indicate that in order to ensure consistently high breakdown fields of tantalum oxide films, oxygen flow rate should be more than the critical flow rate for the increasing flow rate mode.

3.6 References

3.1. S. Schiller, U. Heisig, K. Steinfelder, and J. Strumpfel, Thin Solid Films **63**, 369 (1979).

3.2. S. Schiller, G. Beister, and W. Sieber, Thin Solid Films **111**, 259 (1984).

3.3. K. Chen, M. Nielsen, G. R. Yang, E. J. Rymaszewski, and T.-M Lu, J. Electron. Mater. **26**, 397 (1997).

3.4. J.-Y. Kim, M. Nielsen, E. J. Rymaszewski, and T.-M. Lu, J. Appl. Phys. **87** (3), 1448 (2000).

3.5. M. C. Nielsen, Ph.D. thesis, Rensselaer Polytech. Inst., Troy, NY, 1998.

3.6. S. Berg, H-O. Blom, T. Larsson, and C. Nender, J. Vac. Sci. Technol. A**5**(2), 202 (1987).

3.7. S. Berg, T. Larsson, C. Nender,and H-O. Blom, J. Appl. Phys. **63**(3), 887 (1987).

3.8. S. Berg, H-O. Blom, M. Moradi, T. Larsson, and C. Nender, J. Vac. Sci. Technol. A**7**(3), 1225 (1989).

3.9. H. O. Blom, S. Berg, and T. Larsson, Thin Solid Films **130**, 307 (1985).

3.10. S. Berg, T. Larsson, and H. O. Blom, J. Vac. Sci. Technol. A **4**(3), 594 (1986).

3.11. T. Larsson, H. O. Blom, C. Nender, and S. Berg, J. Vac. Sci. Technol. A **6**(3), 1832 (1988).

3.12. P. Carlsson, C. Nender, H. Barankova, and S. Berg, J. Vac. Sci. Technol. A **11**(4), 1534 (1993).

3.13. L. B. Jonsson, T. Nyberg, and S. Berg, J. Vac. Sci. Technol. A **18**(2), 503 (2000).

3.14. A. H. Eltoukhy, B. R. Natarajan, J. E. Green, and T. L. Barr, Thin Solid Films **69**, 229 (1980).

3.15. G. Lemperiere and J. M. Poitevin, Thin Solid Films **111**, 339 (1984).

3.16.   J. Affinito and R. R. Parsons, J. Vac. Sci. Technol. A2(3), 1275 (1984).

3.17.   F. Shinoki and A. Itoh, J. Appl. Phys. **46**(8), 3381 (1975).

3.18.   VASE®, J.A. Woollam Co., Inc. VASE®.

3.19.   alpha-stepper® 200, Tencor Instruments, part # 035815 Rev. E., 1983 release.

3.20.   4140 pA meter / DC voltage and current source, Yokogawa-Hewlett Packard, Ltd.

3.21.   4280A, 1MHz C Meter / C-V plotter, Yokogawa-Hewlett Packard, Ltd.

3.22.   4192A, LF Impedance Analyzer, 5Hz-13MHz, Yokogawa-Hewlett Packard, Ltd.

3.23.   F. Hmiel, J. Vac. Sci. Technol. A3(3), 592 (1985).

3.24.   C. Chaneliere, S. Four, J. L. Autran, R.A. B. Divine, and N. P. Sandler, J. Appl. Phys. **83**, 4823 (1998).

CHAPTER 4

BREAKDOWN VOLTAGE VERSUS CAPACITANCE DENSITY AND DESIGN LIMITS

Capacitors with the highest (known) capacitance density and highest breakdown voltage are desired in most applications. However, a trade-off exists between these two properties of interest [1-3]. This chapter explores the fundamental basis of this inherent trade-off and establishes the upper limits of the breakdown voltages and capacitance densities that can practically be achieved/targeted using a given dielectric film.

Breakdown voltage and capacitance density of a capacitor are determined by: (a) intrinsic material properties of the dielectric used in the capacitor, i.e., breakdown field and dielectric constant, and (b) structural dimensions of the capacitor, i.e., capacitor area and dielectric thickness. Factors such as electrode-dielectric interface roughness, electrode metal diffusion/reaction into the dielectric, etc., that can further degrade the breakdown of the capacitors.

The breakdown voltage (V_{BR}) and capacitance density (C/A) of a capacitor are related to breakdown field (E_{BR}) and dielectric constant (ε_r), respectively, of the film by

$$V_{BR} = E_{BR}\, d \tag{4.1}$$

$$(C/A) = (\varepsilon_0 \varepsilon_r)/d \tag{4.2}$$

where, d is the dielectric thickness, ε_0 is the permittivity of free space (8.85 x 10^{-12} F/m), ε_r is the dielectric constant of the dielectric film, and A is the area of the capacitor. Hence, in order to achieve the highest possible V_{BR} and (C/A), dielectrics having both high breakdown field and dielectric constant would be the best choice for thin film capacitors. However, data from the literature [3-7] and our own experimental results [1,2] show an inherent trade-off between breakdown field (E_{BR}) and dielectric

69

constant (ε_r) of thin films. This trade-off between E_{BR} and ε_r (both intrinsic properties of the dielectric) is used to determine the design limits for V_{BR} and (C/A) (both are important engineering variables) of the capacitor, with dielectric film thickness and dielectric constant as parameters. The validity of the developed design space is also verified experimentally using Ta_2O_5 films.

4.1   Chapter Overview and Structure

In Sec. 4.2, the interdependency between breakdown field and dielectric constant of thin-film dielectrics is discussed. The empirically determined trade-off between breakdown field and dielectric constant is then translated in terms of suitable engineering variables (breakdown voltage and capacitance density) to develop a design space. This design space can be used as an effective tool to predict the maximum breakdown voltages and capacitance densities that can be achieved with a given dielectric of various thicknesses. In Sec. 4.3, the developed design space is experimentally validated using reactively sputtered tantalum oxide films over a wide range of film thicknesses. The regions of deviation from the "best one can achieve" (BOCA) breakdown voltage using tantalum oxide films are identified and causes for the deviations are also discussed. The summary and conclusions of this chapter are presented in Sec. 4.4.

4.2   Ideal Thin Film Dielectrics: Theoretical Limits

For highest quality films (stoichiometric and free from defects) and homogeneous uniform electric fields, the relationship between maximum dielectric breakdown field, E_{BR} and dielectric constant, ε_r is shown in Fig. 4.1. BOCA line of Fig. 4.1 is represented by equation, E_{BR} (MV/cm) = $\left(20/\sqrt{\varepsilon_r}\right)$, which can equivalently be expressed as $\varepsilon_r E_{BR}^2$ = constant.

70

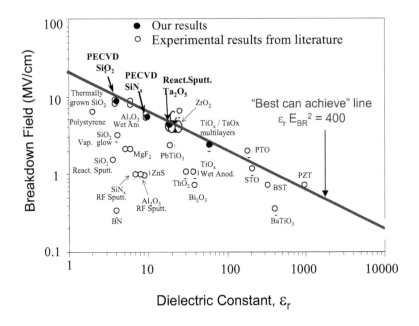

Figure 4.1. Breakdown field as a function of dielectric constant. Breakdown field decreases as dielectric constant increases following $E_{BR}(MV/cm) = 20/(\varepsilon_r)^{1/2}$.

The $\varepsilon_r E_{BR}^2$ = constant implies that the maximum electric energy density that can be stored in a capacitor is constant irrespective of the dielectric material. The fundamental basis for this experimentally observed relation is not yet fully understood.

Blaise and Sarjeant [3,8] explained this trade-off by associating the electric breakdown phenomenon with the detrapping process of the trapped charges in a dielectric film under bias. The charge carriers (electrons/holes) trapped in trapping centers strain the dielectric matrix and thus increase the electromechanical energy near the trapping center. The internal energy concentrated around a trapped charge is released during the detrapping process, producing either local damage (aging) or a catastrophic event (electric breakdown, fracture, etc.). But a minimum strength of electric field is required around the trapping center to detrap it. This minimum electric field is called the "critical detrapping field (E_d)". They argued that high dielectric constant (high-κ) materials have lower critical detrapping fields because of their highly polarizable bond structures. Therefore, there is a higher probability of detrapping of trapped charges in high-κ materials as compared to low-κ materials. In addition, energy released per detrapping (U_p) is also higher in high-κ materials. Thus, lower "E_d" and higher "U_P" of high-κ materials is responsible for their lower breakdown fields compared to that for low-κ materials.

Recently, McPherson *et al.* [6] have also confirmed $E_{BR} \sim 1/\sqrt{\varepsilon_r}$ relationship. They explained this relationship based on a physical model where very high local electric field (in high-κ materials) tends to distort/weaken polar molecular bonds, thereby lowering the enthalpy of activation required for bond breaking by standard boltzman process.

We consider that this relationship, E_{BR} (MV/cm) = $\left(20/\sqrt{\varepsilon_r}\right)$ is an excellent basis for determining the upper limits of the breakdown field of thin film capacitors constituted with a dielectric material of dielectric constant ε_r. To define a design space, which can be used to determine the design limits of the capacitors constituted with the dielectric thin films, the interdependence between fundamental material properties must be translated into suitable engineering variables. Maximum breakdown voltage, V_{BR} and capacitance density, C/A (C is capacitance and A is capacitor area) are the two main figures of merit for any thin film dielectric material to be used in capacitor applications, and hence chosen to be the axes of our proposed design space shown in Fig. 4.2. The thickness, d and dielectric constant, ε_r of the film are the parameters in a proposed design space. Using the empirical correlation for breakdown field and basic fundamentals of capacitance density (Eqs. 4.1 and 4.2), the following relationships are derived:

$$V_{BR} = 1.7 \, \varepsilon_r^{0.5} / (C/A) \qquad\qquad (4.3)$$

$$V_{BR} = 60 \, d^{0.5} / (C/A)^{0.5} \qquad\qquad (4.4)$$

where V_{BR} is in volts, (C/A) is in $\mu F/cm^2$ and d is in μm.

Practical limits to which engineering designers can utilize this design space are dictated by the capability of processing technologies for thin film dielectrics. We call these limits "design boundaries" of the design space. In Fig. 4.2, the proposed design space is shown for the films, with (a) thickness, d, lying between 0.01 to 10 μm, and (b) dielectric constant, ε_r, within 3 to 3000.

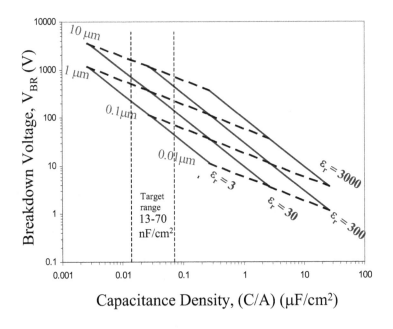

Figure 4.2. **Design space with breakdown voltage and capacitance density as two main figures of merit with film thickness and dielectric constant as parameters. Solid lines are constant ε_r lines and dotted lines are constant thickness lines. The targeted range of capacitance density needed for next generation handheld systems are 13-70 nF/cm² [9,10].**

This design space is an effective and useful tool to identify that (a) what dielectric constant material and (b) what film-thickness, can be used to achieve the required capacitance density at a given operating voltage (determined by specific functional requirement of the capacitor in the circuit). It can also be used to extrapolate and predict whether a given dielectric material can provide a specific combination of breakdown voltage and capacitance density (by reading through line of constant ε_r in Fig. 4.2). Practical capacitance density (using thin film dielectrics) that can be achieved for a given breakdown voltage can also be estimated using this design space.

It is evident from Fig. 4.2 that thicker films are required in order to achieve the same breakdown voltage of films, if materials with higher dielectric constants are used. Also note that the practical capacitance density limit for 1 kV application (in power electronics) is of the order of 5 to 50 nF/cm^2. On the other hand, for less than 10 V applications (signal electronics), capacitance density greater than 1μF/cm^2 can be achieved using 0.1 μm thick films with materials having dielectric constant of more than 100. In general, for an application in the 1 kV range, film thickness must be of the order of 10 μm or greater for films with dielectric constant above approximately 100.

4.3   Design Limits for Capacitors with Ta$_2$O$_5$ Thin Films

In this section, the validity of developed design space (Sec. 4.2) is experimentally verified with reactively sputtered tantalum oxide films over a range of film thicknesses varying from 0.05 to 5.4 μm.

4.3.1   Experimental Methodology

In this study, Ta$_2$O$_5$ thin films were deposited over a range of film thickness (0.05 to 5.4 μm) by pulsed plasma magnetron dc reactive sputtering. The reactive sputtering system is described in chapter 2 (section 2.4). The tantalum target was pre-sputtered in argon plasma for 2 minutes to clean up the target surface and then pre-conditioned for 2 minutes in O_2/Ar plasma before deposition. The sputtering gas used was a mixture of oxygen and argon in the ratio 60:40 by volume. All depositions were carried out at a total pressure of 6 mtorr during sputtering. A 2.5 kW pulsing power was supplied to the target, with 20 kHz voltage pulse, to maintain the plasma during the deposition. No substrate rotation was used during deposition of Ta_2O_5 films.

A metal-oxide-metal (MOM) structure was used for the electrical characterization of these films. Aluminum (0.3 μm thick) was deposited on the wafer as the bottom electrode using electron-beam evaporation. The top aluminum electrode (0.5 mm in diameter and 0.3 μm thick) was also deposited by e-beam evaporation using a shadow mask. The thicknesses of Ta_2O_5 films were measured using a variable-angle spectroscopic ellipsometer, immediately after the films were removed from the deposition chamber. Thickness of the films thicker than 1 μm was also verified from the cross-sectional image of the film obtained from field effect secondary electron microscopy (JSM-6330F, JEOL); ellipsometery does not give accurate film thickness beyond 1 μm.

The Current-Voltage (I-V) characteristics of the Ta_2O_5 films were measured on the MOM structure using Keithley-237 and HP 4140B I-V meter. Standard pin probes were used to make electrical connections to both top and bottom electrodes of the capacitor structure. Voltage was ramped up in steps, with step value of 1V for films less than 1 μm and 5 V for the films thicker than 1μm, until breakdown of the film occurred. Holding time of 1 second was maintained at each voltage step. Electrical breakdown of the film was manifested by the sharp increase in leakage current from 10^{-6} to 10^{-2} A with very small increase in applied voltage. The film was permanently damaged after electrical breakdown. The capacitance measurements were taken on HP 4280 C-V meter and HP 4192 impedance meter at 1 MHz with 30mV rms signal level.

4.3.2 Experimental Results and Discussion

Tantalum oxide films of different thickness ranging from 0.05 to 5.4 μm were deposited and characterized. Four to five films of each thickness with 15-20 capacitors per film were tested to confirm the repeatability of the electrical data. The breakdown voltage and capacitance density for tantalum oxide films are shown in Fig. 4.3. The solid line in Fig. 4.3 is extracted from the design space equation (Eq. 4.3 or Fig. 4.2) for $\varepsilon_r = 25$. Amorphous Ta_2O_5 have dielectric constant around 23-25 [11-16]. The solid line represents the BOCA combinations of breakdown voltages and capacitance densities that can be achieved using amorphous Ta_2O_5.

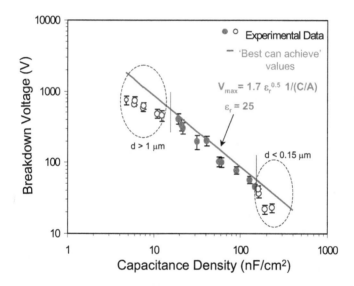

Figure. 4.3. Breakdown voltage as a function of capacitance density for tantalum oxide thin films. The breakdown voltage increases from 10V to 900V and capacitance density drops from 150 to 7 nF/cm², when film-thickness is increased from 0.05 to 5.4 μm. Solid line (—) shows the "best can achieve" values, solid dots (•) are the experimental values that are in agreement with the "best can achieve" values, and hollow dots (o) are the regions of significant deviations from the "best can achieve" values.

Three regions can be identified in Fig. 4.3. Experimental results with films within 0.15 to 1 μm range are in close agreement with the BOCA values of breakdown voltages and capacitance densities using Ta_2O_5. Films thinner than around 0.15 μm and that thicker than around 1 μm exhibit relatively lower breakdown voltages compared to BOCA values.

In this chapter, as a convention, films thicker than ~1 μm are referred as "thicker" films and films thinner than ~0.15 μm are referred as "thinner" films. Remaining sections of this chapter discuss the experimental determination of the prime causes of relatively lower electrical breakdown of both thicker and thinner films. The measures to bring the breakdown voltages of both thicker and thinner films to their BOCA values are also presented.

4.4   Deviation from BOCA Performance

In this section, the term "film thickness" is referred to nominal thickness of the film, which may or may not have stoichiometric composition uniformly across it. The films with thickness less than 1 μm are also referred to "submicron" films. Unless stated, the electrical properties reported in this section are for films deposited without intentional substrate cooling.

4.4.1   Breakdown Field and Breakdown Voltage

Ideally, the breakdown field is an intrinsic material property and is expected to be constant with film-thickness. The breakdown field of tantalum oxide films as a function of nominal film-thickness is shown in Fig. 4.4. Breakdown fields of the films with thickness between ~ 0.15 to 1 μm are nearly constant around 3 to 3.5 MV/cm (shown in solid dots, •). The breakdown field then deteriorates from 3.0 to 1.5 MV/cm with further increase in film thickness from 1 μm to 5.4 μm.

77

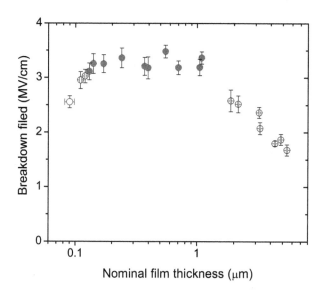

Figure 4.4. Breakdown field of tantalum oxide films as a function of film thickness, showing a decline in breakdown field for "thinner" films (< ~0.15 μm) and also for "thicker" films (>~ 1 μm).

The breakdown voltage is plotted against nominal film thickness in Fig. 4.5. For "thinner" films (< 0.15 μm), the rate of increase in breakdown voltage increases sharply with film thickness. For films with thickness around 0.15 to 1 μm, the breakdown voltage of the tantalum oxide films increases linearly, roughly from 45 V to 350 V with film thickness. The rate of increase of breakdown voltage with film thickness then tapers down beyond 1 μm for "thicker" films. The thickest films deposited in this work (~ 5.4 μm) showed an average breakdown voltage around 800 V.

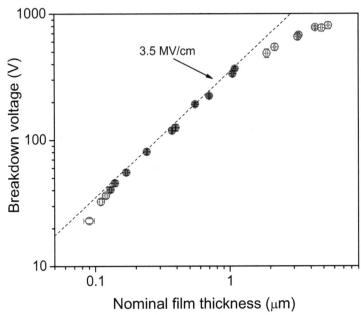

Figure 4.5. Variation of breakdown voltage with film thickness, showing the deviation from the straight line corresponding to 3.5 MV/cm, for films thicker than 1 μm, and for films thinner than around 0.15 μm.

The distribution of breakdown voltage for different film thicknesses is shown in Fig. 4.6. The cumulative probability of failure (y-axis), below a particular breakdown voltage (x-axis), is plotted in Fig. 4.6. A tighter distribution around a given breakdown voltage is an indicative of reliable dielectric films around that operating voltage range. The distribution gets wider with film thickness. Thickest films deposited in this work (~5.4 μm) showed a very wide distribution in breakdown voltage.

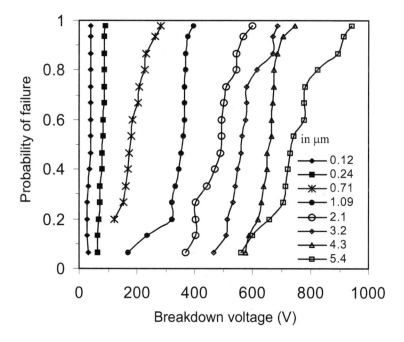

Figure 4.6. **Cumulative probability of failure due to electrical breakdown (y-axis) below a particular breakdown voltage (x-axis) for films with thickness from 0.12 to 5.4 μm.**

4.4.2   Dielectric Constant and Dielectric Loss

The variation of measured dielectric constant of tantalum oxide films with film-thickness is shown in Fig. 4.7(a). The measured dielectric constant of the film is obtained from capacitance measurements. The measured dielectric constant of the film increases from 23 to 29 when the film thickness is increased from 0.05 to 5.4 μm. The measured dielectric constants for "submicron films" remain around 23 to 25. The measured dielectric constants for "submicron films" are consistent with other studies in the literature reporting dielectric constant around 23 to 25 for stoichiometric and amorphous Ta_2O_5 films [11-16].

The measured dielectric loss at 1 MHz is plotted against film thickness in Fig. 4.7(b). "Submicron films" show a low dielectric loss (< 0.02), while the films thicker than 1 μm shows high dielectric losses (> 0.08).

It is important to note that both "thinner" (< 0.15 μm) and "thicker" (> 1 μm) films show relatively lower breakdown voltages (Fig. 4.5), but it is only "thicker" films that also show higher dielectric loss and increase in measured dielectric constant. Metals have infinite dielectric constant and dielectric loss, and therefore "metallic" content in the dielectric films will cause simultaneous increase in measured dielectric constant and dielectric loss. Hence, the presence of sub-stoichiometric oxide (either due to the presence of Ta rich or oxygen deficient oxides) is indicated in "thicker" films. The presence of sub-stoichiometric oxides will cause lower breakdown fields and higher leakage current densities. In contrast, measured dielectric constant and dielectric loss results do not indicate any "metallic" content in "thinner" films. Therefore, the cause of lower breakdown voltages for "thinner" films appears to be different from that for "thicker" films.

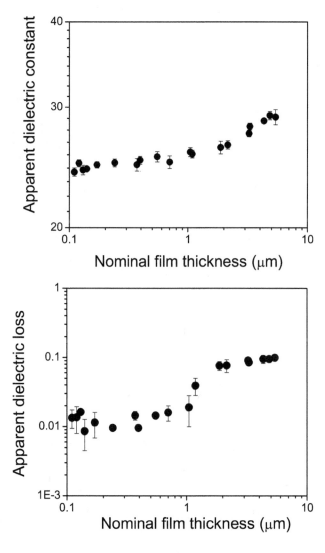

Figure 4.7. (a) Variation of measured dielectric constant with film thickness. (b) Measured dielectric loss versus film thickness, showing relatively higher loss for films thicker than around 1 μm.

4.4.3 Leakage Current Density

Typical variation of the leakage current density (J) with applied electric field (E) for 7 different tantalum oxide films (0.12 to 5.4 μm in thicknesses), are shown in Fig. 4.8.

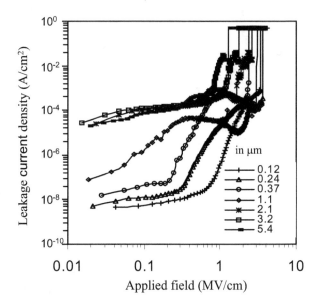

Figure 4.8.   Dependence of leakage current density on applied field for films with thickness varying from 0.12 to 5.4 μm.

The leakage current densities for submicron films (as shown for 0.12, 0.24 and 0.37 μm) can be broadly classified into three main regimes: ohmic, nonlinear and breakdown regime. The ohmic regime exists at low electric fields (< ~0.2 MV/cm), where leakage current density increases linearly with applied electric field. This ohmic conduction is attributed to thermal excitation or hopping of trapped electrons from one trapped site to another trapped site. This region is followed by a nonlinear conduction regime (around 0.2 to 1 MV/cm). For a given film-thickness, the magnitude of leakage current density increases by two to three orders of magnitude in nonlinear conduction regime as compared to the ohmic regime. The conduction mechanism in the nonlinear conduction regime is a combination of various bulk and electrode limited conduction mechanisms; e.g., Schottky barrier limited, Poole-Frenkel emission, and space charge limited conduction mechanisms [17]. The leakage current in both ohmic and nonlinear regions is reversible in nature with respect to applied electric field. At still higher fields (~1.5 to 3 MV/cm), the leakage current increases sharply with applied field until film breaks down irreversibly. Films after the breakdown were found to be permanently electrically shorted and showed leakage currents above the current compliance limit (10^{-2} A), even at very low voltages during the repeated scan on the same structure. In submicron films, the leakage current density at a given applied field increases with film thickness in both ohmic as well as nonlinear regime.

Unlike submicron films, films thicker than around 1 μm showed very high leakage currents even at very low applied electric field. This high value of leakage current density is nearly constant over the entire range of voltage scan. No major effect of film thickness on leakage characteristics was observed, once the films were grown over ~ 2 μm. Typical leakage current densities for thicker films (>1 μm) were at least 3 orders of magnitude higher than that for "submicron" films.

4.4.4 Chemical Composition

X-Ray photoelectron spectroscopy (XPS) is used to determine the chemical composition of the top surface of three different films (0.12, 1.2 and 5.4 μm thick). In this XPS analysis, 0.12 μm thick film represents "thinner" films, 5.4 μm thick film represents "thicker" films, and 1.2 μm thick film represents the transition where breakdown field starts decreasing. XPS scans (0-600 eV) are shown in Fig. 4.9. These scans have been offset along the y-axis for the sake of clarity. All of the tantalum peaks (4f, 4d, 4p, 4s) are present at their respective positions. Tantalum $4f_{7/2}$ and $4f_{5/2}$ peaks are present at 28.4 and 26.5 eV. Oxygen 1s peak is present at 530.9 eV. These peak positions are in close agreement with the values reported in literature [18].

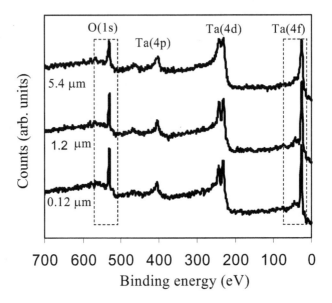

Figure 4.9. **Wide range XPS scan, indicating the presence of various tantalum and oxygen peaks for films with three different thickness (0.12, 1.2 and 5.4 μm).**

The close-up views of O(1s) and Ta(4f$_{7/2}$ and 4f$_{5/2}$) are shown in Fig. 4.10(a) and 4.10(b), respectively. It is evident that peaks corresponding to Ta(4f$_{7/2}$ and 4f$_{5/2}$) do not change appreciably, but the peaks corresponding to O(1s) signal decreased significantly, when the film thickness is increased from 0.12 to 5.4 μm. The ratio of the areas under the O(1s) peak and Ta(4f$_{7/2}$ and 4f$_{5/2}$ combined) peak is calculated and adjusted according to the sensitivity factors for oxygen and tantalum signals.

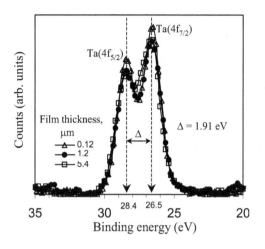

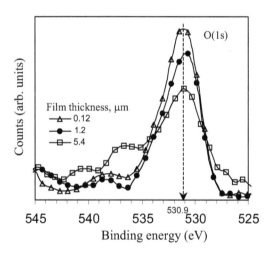

Figure 4.10. High resolution XPS scans showing **(a)** nearly the same area under tantalum 4f peaks and **(b)** relative lowering in the area under oxygen 1s peaks, with increase in film thickness from 0.12 to 5.4 μm.

This calculated ratio gives an estimate of the stoichiometry ("x" in TaO_x) of the top surface of a given film. The XPS results are summarized in Table 4.1 and potted in Fig. 4.11. The oxygen to tantalum ratio in the deposited film's top surface decreases from 2.60 to 1.38 when film thickness was increased from 0.12 to 5.4 μm. This shows that tantalum oxide films get sub-stoichiometric or tantalum rich (or oxygen deficient) with deposition time. The oxygen to tantalum ratio ("x" in TaO_x) should be around 2.5 for stoichiometric tantalum oxide films [11].

Table 4.1. Atomic percent of tantalum and oxygen present in reactively sputtered tantalum oxide films of different film thicknesses

Film Thickness (μm)	Atomic %		Stoichiometry (x in TaOx)
	Ta %	O %	
0.12	25	65	2.60
1.09	36	64	1.77
5.4	42	58	1.38

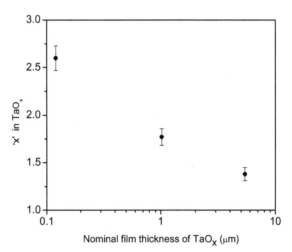

Figure 4.11. Stoichiometric ratios ("x" in TaO_x) of the films deposited in our reactive sputtering system with thickness varying from 0.12 to 5.4 μm.

Thus, the "thinner" films are stoichiometric in chemical composition. But, as film grows, it starts becoming more and more oxygen deficient (or Ta rich). Therefore, "thicker" films have graded chemical composition, with top few layers highly sub-stoichiometric. The presence of sub-stoichiometric oxide appears to be responsible for relatively lower breakdown fields, apparently higher dielectric constants along with higher dielectric losses, and higher leakage current densities of the "thicker" films.

In the next section, the prime cause for the evolution of sub-stoichiometric oxides with deposition time is discussed. Increase in substrate temperature during deposition is correlated with the deposition of sub-stoichiometric oxide.

4.4.5 Control of Film Stoichiometry

In this section, the evolution of film stoichiometry, and hence the electrical properties of tantalum oxide films are compared for two different schemes of depositions; (1) when no substrate cooling was provided and the substrate temperature was allowed to increase with deposition time, and (2) the other when the substrate temperature was kept below 40°C by water cooling.

4.4.5.1 Substrate Temperature

The temperature of the substrate during sputtering was monitored as a function of time during the film growth and is plotted in Fig. 4.12 for two cases: (a) with no substrate cooling and (b) when substrate was water-cooled. In both cases, 2.5 kW of sputtering power was used during sputtering.

When no substrate cooling is provided during deposition, the temperature of the film increases with time because of the transfer of kinetic energy from continuously bombarding sputtered species to the depositing film. The temperature of the substrate increases nearly linearly with deposition time during first 20 to 25 minutes (time

88

required to deposit ~1 μm). After that, the rate of increase of temperature with deposition time tapers down and saturates at around 220°C, once the rate of heat gain due to the bombarding species balances with the rate of heat loss to the surroundings. In contrast, the substrate temperature remains below 40°C when the substrate is water-cooled.

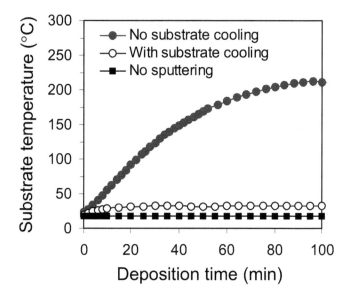

Figure 4.12. Substrate temperature with deposition time for the cooled substrate and non-cooled substrate cases.

4.4.5.2 Oxygen Partial Pressure

The oxygen partial pressure in the sputtering chamber is plotted against deposition time in Fig. 4.13. The total oxygen flow rate was kept constant at 8.0 sccm during the deposition. At 0.0 kW, there would be no sputtering at the tantalum target. Hence, the oxygen partial pressure remains constant at 10.1 mtorr at 0.0 kW. With no substrate cooling, at 2.5 kW sputtering power, the oxygen partial pressure first remains nearly constant until around 25 minutes (deposition time for ~1 μm thick film) and then increases with deposition time.

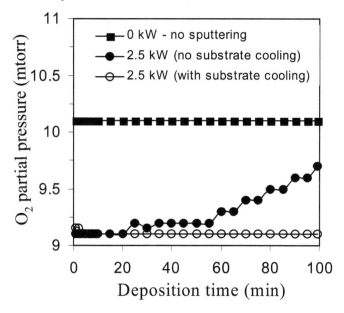

Figure 4.13. **Variation of partial pressure of oxygen in the deposition chamber as a function of deposition time for 0.0 and 2.5 kW sputtering power.**

The difference in partial pressures at 0.0 and 2.5 kW cases gives an estimate of the amount of oxygen used in oxidizing tantalum to form tantalum oxide. This indicates that the amount of oxygen consumed in oxidizing tantalum to tantalum-oxide goes down with deposition time and causes the formation of sub-stoichiometric oxide, if substrate cooling is not provided. On the other hand, when substrate is water-cooled during deposition, the partial pressure of oxygen remained almost constant during the whole deposition and therefore, the amount of O_2 consumed during film deposition does not decrease with time.

4.4.5.3 Stoichiometric Oxide with Substrate Cooling

The stoichiometric ratio ("x" in TaO_x) is compared in Fig. 4.14 for films deposited (a) with substrate cooling and (b) without substrate cooling. Films deposited with substrate cooling are stoichiometric in chemical composition over the whole range of film thickness (till 5.4 μm). Therefore, we conclude that by providing substrate cooling during deposition, stoichiometric oxides can be deposited, even for "thicker" films. The possible mechanism responsible for the improvement in stoichiometry due to substrate cooling during reactive sputtering is described in the next paragraph.

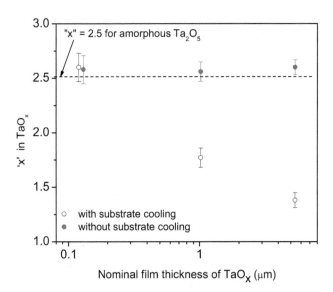

Figure 4.14. Stoichiometric ratios ("x" in TaO_x) for the films deposited with and without substrate cooling.

Oxygen in the chamber reacts with Ta target to form stoichiometric tantalum oxide, i.e., Ta_2O_5. When the sputtering ions (Ar or O ions) bombard the target, the oxide on the target can sputter off as different TaO_x species (TaO, Ta_2O_3, etc.). These TaO_x

species, when reaches the substrate, have to react with O_2 to form Ta_2O_5 at the substrate. For this reaction to occur, first O_2 has to adsorb on the substrate and then react with TaO_x species to form Ta_2O_5. At higher temperatures, O_2 adsorption decreases due to reduction in sticking coefficient at higher temperatures. Because substrate temperature increases with deposition time, lesser amount of oxygen adsorbs than required for stoichiometric oxide (Fig.4.13). Hence, once substrate temperature is kept below 40 °C by providing substrate cooling, stoichiometric oxides could be formed.

4.4.6   Improved Electrical Properties with Substrate Cooling

As shown in the previous sections, the growth of sub-stoichiometric oxide is responsible for the relatively lower breakdown voltage, higher apparent dielectric constant and higher leakage current density of "thicker" films. As evident from Fig. 4.4, the breakdown field remains nearly constant around 3 to 3.5 MV/cm till around 1 μm of film thickness. But once films start getting oxygen deficient above 1 μm, the breakdown field decreases from 3.5 to 1.5 MV/cm. The presence of un-oxidized tantalum in thicker films is responsible for the increase in apparent dielectric constant, as shown in Fig. 4.7(a). Although, the presence of metallic content leads to the increase in dielectric constant of the film, dielectric loss also goes up. Fig. 4.7(b) shows that the dielectric loss also goes up by a factor of 4 for thicker films. The presence of both high dielectric constant and high dielectric loss simultaneously confirms the presence of unoxidized tantalum in the film. Since the thicker films are mostly metallic across the film cross-section, the effective electric field is much higher than the applied field and hence gives higher leakage current density at a given field.

Superior breakdown and leakage properties of the TaO_x films are obtained when the substrate is water cooled during deposition compared to when there is no substrate cooling. The effect of substrate cooling on leakage current density versus applied field plot is shown in Fig.4.15 for the thickest film of thickness around 5.4 μm. Leakage current density improves by around three orders of magnitude at low applied fields (up to 0.1 MV/cm) due to substrate cooling.

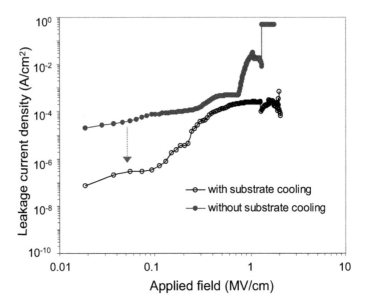

Figure. 4.15. Effect of substrate cooling on leakage current density of 5.4 μm film.

Distribution of breakdown voltage for 5.4 μm films, as shown in Fig. 4.16 (a) also shifts from around 700 to 1350 V. Effect of substrate cooling on dielectric constant and dielectric loss is shown in Fig. 4.16(b). Dielectric constant decreases to around 25, which is same as the value reported in literature for stoichiometric Ta_2O_5 films. Dielectric loss also decreases by factor 8 due to substrate cooling.

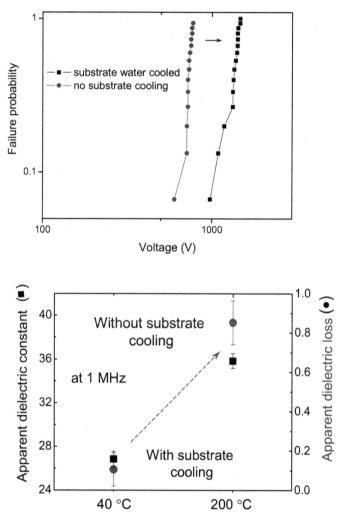

Figure 4.16. Effect of substrate cooling on (a) distribution of breakdown voltage and (b) dielectric constant and dielectric loss of 5.4 μm films.

Hence, we conclude that substrate heating, during deposition of "thicker" films (> around 1 μm), leads to deposition of sub-stoichiometric TaO_x, and therefore causes lower breakdown voltage, higher leakage current density and high dielectric losses. Substrate cooling during deposition improves the breakdown and leakage properties of "thicker" Ta_2O_5 films. The breakdown voltage of the films deposited with substrate cooling now closely follows the BOCA value (as shown in Fig. 4.17). Now, the only region of deviation from BOCA performance is corresponding to "thinner" films (< ~0.15 μm). The prime cause of deviation for "thinner" films is discussed in the next section.

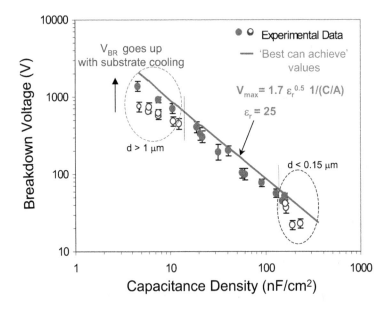

Figure 4.17. Breakdown voltages of the "thicker" films deposited with substrate cooling closely follows the predicted "best can achieve" values. Lower breakdown voltages of "thinner" films is discussed in the next section.

4.4.7   Lower Breakdown Fields in Thinner Films (< 0.15 μm)

The lower breakdown fields for "thinner" films (< 0.15 μm) are due to micro-roughness of the bottom electrode, which can be a significant fraction of the total film-thickness, as shown in Fig. 4.18.

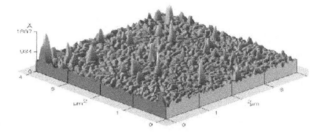

Figure 4.18.   AFM image of the aluminum bottom electrode (~0.3 μm), deposited using e-beam evaporation on silicon wafer. The presence of sharp peaks (~ 0.1μm) of bottom aluminum electrode are responsible for the relatively lower electrical breakdown of the thin films (< 0.4 μm).

The micro-roughness of the bottom electrode increases the effective electric field at the sharp and pointed regions of the bottom electrode and results into a premature breakdown of the thin films at these locations. The increase in electric fields at the sharp and pointed features is known as "lightening rod" field amplification (shown in Fig. 4.19).

'lightening rod' field amplification

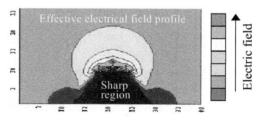

Figure 4.19.   Higher effective fields at the sharp and pointed regions due to 'lightening rod' field amplification.

The increase in local electric field at sharp peaks in bottom electrodes causes cratering and burning out of the top metal electrode, leading to local breakdown, as shown in Fig. 4.20(a). A second sweep of the I-V characteristic indicates an absence of the local breakdowns, as shown in Fig. 4.20(b). The presence of pinholes or defects in the films can also result in the shorting of top and bottom electrodes, especially for relatively thin films.

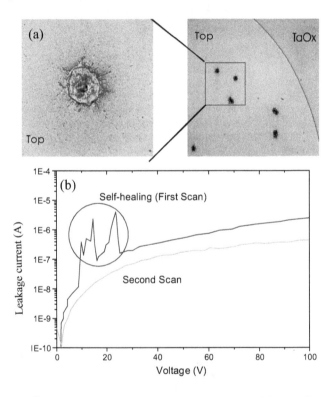

Figure 4.20. **(a) Photographs showing the burnt out areas of the top electrode of films thinner than 0.4 μm, leading to premature breakdown. (b) Films sometimes self-heal during the first voltage scan and show a smooth current-voltage curve in the second scan of the same film.**

The following set of experiments was performed to evaluate the effect of bottom electrode roughness on the breakdown fields of thinner films. Tantalum oxide films (~0.14 μm) were deposited on different aluminum electrodes of varied thickness. The thickness of the bottom aluminum electrode was varied from 0.05 to 0.6 μm in order to get different bottom electrode roughness. Fig. 4.21 shows the representative AFM scans of these aluminum electrodes of different thickness.

Z dimension : 10 X

t: thickness of Aluminum bottom electrode

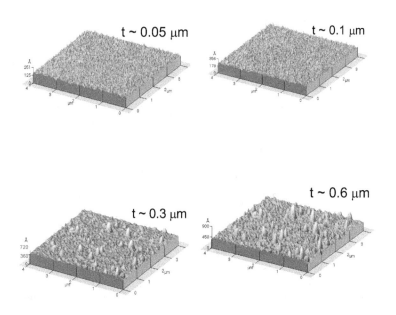

Figure 4.21. **Representative AFM scans of the aluminum bottom electrodes of variable thickness (0.05 to 0.6 μm).**

The root mean square (rms) and peak-to-valley (P-V) roughness of aluminum bottom electrodes of variable thickness is shown in Fig. 4.22. Although rms roughness does not vary significantly with bottom electrode thickness, P-V roughness increases with aluminum film thickness. For aluminum thickness ~ 0.3 μm, the P-V roughness is comparable to the tantalum oxide film thickness and hence causes premature electrical breakdown at these sharp points.

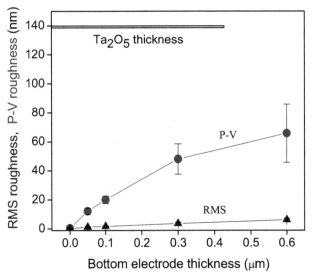

Figure 4.22.   RMS and P-V roughness of bottom aluminum electrodes

MIM capacitors were made using above-mentioned aluminum bottom electrodes and tested for breakdown fields. Fig. 4.23(a) summarizes the improvement in breakdown fields as P-V roughness of the bottom electrodes goes down. The breakdown fields first improve linearly as P-V roughness decreases from 65 nm to 15 nm. But once the P-V roughness becomes less than 10% of the dielectric film thickness, breakdown field for Ta_2O_5 films approaches the BOCA value. Hence, BOCA breakdown voltages can be achieved, even for films thinner than 0.15 μm, by keeping the interface roughness less than 10% of dielectric thickness. This improvement in breakdown voltage for "thinner" films is shown in Fig. 4.21(b).

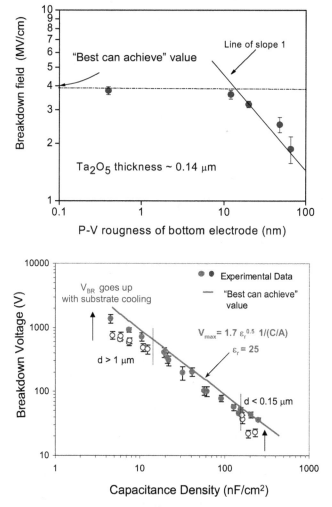

Figure 4.21. (a) Improvement in breakdown field as P-V roughness of the bottom interface decreases, and (b) 'Best can achieve' breakdown voltages are achieved for thinner films by controlling the bottom interface roughness.

4.5   Summary and Conclusions

A correlation between electrical breakdown field and dielectric constant, E_{BR} (MV/cm) $= 20/(\varepsilon_r)^{1/2}$, is empirically determined. Consequently, a design space in terms of breakdown voltage and capacitance density, with film thickness and material dielectric constant as parameters, is developed. This design space sets the limits for BOCA breakdown voltages and capacitance densities using a particular dielectric. The validity of the developed design space is experimentally verified with Ta_2O_5 thin films over a wide range of film thickness (0.05 to 5.4 μm). Detractors causing the deviations from the BOCA breakdown voltages were identified and corrected. Lower breakdown field for the thin films (< 0.15 μm) is caused by the increase in local electric field due to the micro-roughness of the bottom electrode. Substrate heating during film deposition is the prime cause of the relatively lower breakdown of the thicker films (> 1 μm). Consequently, substrate cooling improves the breakdown voltage of thicker films to their BOCA values.

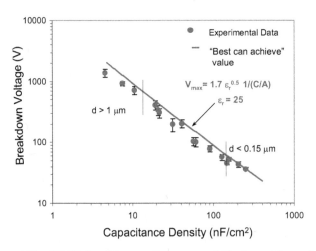

Figure 4.22.   BOCA breakdown voltages are achieved over the entire range of film thickness (0.05 to 5.4 μm)

Thus, we conclude the following:

- Based on empirically determined relationship, E_{BR} (MV/cm) $= \left(20/\sqrt{\varepsilon_r}\right)$, the total amount of electrostatic energy that can be stored in any thin film dielectric material is constant.

- The developed "best one can achieve" (BOCA) design space serves as an effective engineering tool to guide (a) what dielectric constant material and (b) of what film-thickness, is needed to achieve the required capacitance at a given operating voltage. This design space depicts an aggressive trade-off between breakdown field and dielectric constant and like other designs, a safety margin should be provided to come up to final design value.

- Substrate heating during deposition of thicker films (>1 μm) leads to sub-stoichiometric tantalum oxide causing lower breakdown and higher leakage in thicker films.

- Micro-roughness of the bottom electrode degrades the breakdown fields of thin dielectric films. Electrode P-V roughness should be less than 10% of dielectric film thickness to avoid premature breakdown of dielectric films.

4.6 References

4.1   P. Jain and E. J. Rymaszewski, IEEE Trans. Adv. Packag., **25**, 454 (2002).

4.2   P. Jain, J. Y. Kim, Y. Xiao, R. Natarajan, E. J. Rymaszewski, R. J. Gutmann, and T. P. Chow, *Proc. CPES annual review* 2000, Blacksburg, VA, pp. 155-158, 2000.

4.3   Blaise and W. J. Sarjeant, "in Handbook of Low and High Dielectric Constant Materials and Their Applications," edited by H. S. Nalwa, 1st ed., vol. 2, Academic Press, NY, 1999, pp. 138-188.

4.4   E. J. Rymaszewski, "in *ASM Handbook, Material Selection and Design*," edited by G.E. Dieter (chief Ed.) ASM International, Materials Park 1997, vol. 20, Sec. 6, pp. 615-621.

4.5   L. I. Maissel, and R. Glang, "*Handbook of Thin Film Technology*," 1st ed. (reissue), McGRAW-HILL book Co., 1983, ch.16, pp. 21-33.

4.6   J. McPherson, J. Kim, A. Shanware, H. Mogul, and J. Rodriguer, *Proc. IEDM 2002,* San Francisco, CA, pp. 633-635. 2002.

4.7   J. Y. Kim, Ph.D. thesis, Rensselaer Polytech. Inst., Troy, NY, 1998.

4.8   G. Blaise and W. J. Sarjeant, J. Appl. Phys. **77**, 2916 (1995).

4.9   International Technology Roadmap for Semiconductors, Assembly and Packaging, 2001 Ed., http://public.itrs.net/files/1999_SIA_Roadmap/Assembly.pdf.

4.10   P. Chahal, R. R. Tummala, G. Allen, and M. Swaminathan, IEEE Trans. Comp., Packag., Manufact. Technol. B. **21**, 184 (1998).

4.11   Chaneliere, S. Four, J. L. Autran, R.A. B. Divine, and N. P. Sandler, J. Appl. Phys. **83**, 4823 (1998).

4.12   K. Chen, M. Nielsen, S. Soss, E. J. Rymaszewski, T.-M. Lu, and C. Wan, IEEE Trans. Comp., Packag., Manufact. Technol. B **20**, 117 (1997).

4.13   M. C. Nielsen, Ph.D. thesis, Rensselaer Polytech. Inst., Troy, NY, 1998.

4.14   S.-D. Cho and K.-W. Paik, Mater. Sci. Eng. B. **67**, 108 (1999).

4.15   X. M. Wu, P. K. Wu, T. –M. Lu, and E. J. Rymaszewski, Appl. Phys. Lett. **62**(25), 3364 (1993).

4.16   J.-Y. Kim, A. Garg, E. J. Rymaszewski, and T.-M. Lu, IEEE Trans. Comp. Packag. Technol. **24**(3), 526 (2001).

4.17   D. R. Lamb and D. Robert, "*Electrical conduction mechanisms in thin insulating films*," (Methuen, London 1967), Chap. 1-6.

4.18   J. F. Moulder, W.F. Stickle, P.E. Sobol, and K. D. Bomben, "in *Handbook of X-Ray Photoelectron Spectroscopy*", edited by J. Chastain (Perkin Elmer Corp., 1992).

CHAPTER 5

ELECTRICAL CHARACTERIZATION/ EVALUATION OF THIN FILM CAPACITORS

The control of power distribution (supply) noise is becoming increasingly important as the performance of packaged electronics is increasing. Decoupling capacitors are provided as close to the switching circuits as possible to contain power distribution noise within acceptable limits. This chapter discusses the electrical properties and performance of thin film decoupling capacitors, based on dielectrics compatible with semiconductor processing. In this chapter, the performance advantages offered by embedded thin film decoupling capacitors over discrete capacitors are quantified, with experimental emphasis on Ta_2O_5 decoupling capacitors. The performance of Ta_2O_5 decoupling capacitors is also compared with SiO_2 and Si_3N_4 decoupling capacitors.

5.1 Chapter Overview and Structure

In this chapter, the electrical properties and performance of Ta_2O_5 thin film capacitors are presented over a range of operating frequencies (dc to 20 GHz). The electrical properties evaluated in this chapter are dielectric constant, dielectric dispersion, dielectric loss, breakdown field, and leakage current density. These electrical properties are also tested for SiO_2 and Si_3N_4, using same design test vehicles and similar test conditions. SiO_2 and Si_3N_4, being the standard dielectrics used in microelectronics industry, serve as benchmarks to evaluate the potential of Ta_2O_5 films for decoupling capacitors.

In section 5.2, the role of decoupling capacitors to control power distribution noise is discussed. Section 5.3 describes the design and fabrication of the test vehicles, used for electrical characterization over a range of frequencies (dc to 20 GHz). Section 5.4 presents the experimental methodology used for the electrical characterization of

Ta_2O_5, SiO_2 and Si_3N_4 thin film capacitors. The key results are reported and discussed in section 5.5, where the dielectric constants and dielectric losses of Ta_2O_5, SiO_2 and Si_3N_4 are reported up to 20 GHz. The self-inductances of power connect vias are determined from scattering parameter measurements. The dc leakage current densities and breakdown fields of thin films of these materials are also compared. The results presented in this section provide quantitative estimates of the leverage offered by thin film embedded capacitor technology for power decoupling, based on dielectrics compatible with semiconductor processing. The summary and conclusions of this chapter are presented in section 5.6.

5.2 Thin Film Decoupling Capacitors

The power distribution noise, also know as ΔI noise, is generated due to the simultaneous switching of numerous drivers in digital or mixed signal application circuits. ΔI noise gives rise to the voltage fluctuations across the power supply buses, and can also couple through quiet logic circuit drivers and appear as spurious voltage signals on the input terminals of logic receiver circuits and sometimes may cause erroneous switching of the circuits [1-7]. The cause of generation of ΔI noise is explained using Fig. 5.1.

In digital circuits, the input node of the transistors can be represented by a capacitor connected in series with on-chip resistance (as shown in Fig. 5.1). For instance in CMOS transistors, the MOS capacitance (on-chip load) has to be charged and discharged to switch binary states. This could happen once every 1.0 ns for a 1 GHz high performance microprocessor. Therefore, the power supply needs to provide a large amount of transient current to the MOS capacitance switching from low-to-high voltages, and the ground terminals need to sink a large amount of current from the capacitance switching from high-to-low voltages.

106

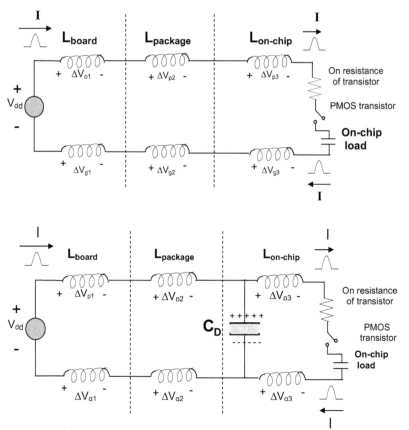

Figure 5.1. **(a) Power supply noise generated due to current transients flowing through board, package, and chip level inductances. (b) Decoupling capacitor acts as temporary power supply near the switching transistors. Board and package level inductances can be bypassed by providing decoupling capacitors at on-chip level.**

The transient currents, that flow through the inductance of the board (PWB), package, and on-chip voltage/ground distribution wirings/planes, cause voltage fluctuations (ΔV) on the power supply rails of the chip, as described by

$$\Delta V = L_{eff} \left(di/dt \right) \tag{5.1}$$

where di/dt is the current transient and L_{eff} is the total effective inductance, given by

$$L_{eff} = L_{board} + L_{package} + L_{on-chip} \tag{5.2}$$

One effective way to suppress the power distribution noise described above is to provide decoupling capacitors (C_D) close to the switching circuits to supply the current needed by them during switching (as shown in Fig. 5.1(b)). Decoupling capacitors reduce power supply level fluctuations by charging up during the steady state and then by supplying the instantaneous current required during simultaneous circuit switching. Thus, they act as temporary source of charge physically close to the switching circuits.

The power distribution system for typical large high-speed machines is diagrammed in Fig. 5.2. A regulated voltage supply feeds current into the frame-distribution system, which is an assembly of laminar bus bars. At frame levels a large amount of decoupling capacitances (SMT discrete capacitors) are inserted for filtering the lower frequency component of the (di/dt). Typically, regulated power supplies have in built decoupling capacitors in them. From this point, current flows through the board bus into the board planes, where mid frequency decoupling capacitors are mounted on the boards. As power leaves the board, it enters the module through pins. Current within

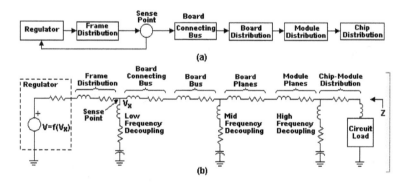

Figure 5.2.   Power-distribution system: (a) Block diagram, and (b) Equivalent circuit

108

the module is further distributed on near solid planes into the vertical vias that are connected to the chip pads. Finally, current reaches the switching transistors through on-chip multilevel metallization (MLM) of the IC. Therefore, a good place to mount high-frequency decoupling capacitors is at the module/IC interface or within the MLM of the ICs. Both low and mid frequency decoupling capacitors are always provided in typical electronic systems. High-frequency on-chip decoupling capacitors need further research and development in terms of design and integration, especially for the electronic systems operating at GHz clock frequencies.

Thus, on-chip decoupling capacitors decouple the instantaneous current flow during circuit switching from the board (PWB) and package inductances. Hence, by providing a decoupling capacitor as shown in Fig. 5.1(b), the total voltage fluctuation (ΔV) becomes

$$\Delta V = \Delta V_{cap} + \Delta V_{p3} \qquad (5.3)$$

where, ΔV_{p3} is the voltage drop at on-chip interconnect inductances and ΔV_{cap} is the voltage drop across the decoupling capacitor. ΔV_{cap} is given by [8]

$$\Delta V_{cap} = Q/C + V_{ci} + L_{cap}\left(di/dt\right) + Ri \qquad (5.4)$$

where, Q is the charge on the capacitor, C is the equivalent series capacitance (ESC) for an equivalent capacitor having an ideal dielectric, V_{ci} is the voltage drop caused by dielectric dispersion, (di/dt) is the current transient, L_{cap} is the series inductance, and R is the equivalent series resistance (ESR) of the decoupling capacitor.

Thin film decoupling capacitors can be integrated within the IC chip or packaging module substrate underneath the IC chip or in PWB. By using the embedded thin film capacitor approach, not only compact packaging modules can be realized, but parasitic inductances can be significantly reduced (by 2 to 3 orders of magnitude) due to shorter connections and elimination of leads. The performance of these capacitors is

109

determined by the intrinsic electrical properties of thin film dielectrics used in them, and hence, the evaluation of electrical properties of thin film dielectrics is required.

5.3 Test Vehicle: Design and Fabrication

This section describes the design and fabrication of the coplanar test vehicle used for the electrical characterization of Ta_2O_5, SiO_2 and Si_3N_4 thin film capacitors. The mask design

5.3.1 Test Vehicle Design

Fig. 5.3 shows an overall arrangement of coplanar microwave probes on the designed and fabricated capacitor structures. Two ground-signal-ground (GSG) probes are shown, measuring the S-parameter measurements from 45 MHz to 20 GHz. The details of this high frequency characterization technique are discussed in the next section. The top and cross-sectional views of Ta_2O_5, SiO_2, and Si_3N_4 capacitors, used for broadband (dc to 20 GHz) electrical characterization, is shown in Fig. 5.4(a). Series equivalent circuit model for the capacitor is also shown in Fig. 5.4(a).

The test vehicle consists of parallel plate capacitor structures in metal-insulator-metal (MIM) configurations. The fabrication steps of these capacitors are discussed in the next section. The test vehicle has six different size capacitors, with the dimensions as listed in Table 5.1. The via holes are used to bring up the electrical contact from the bottom metal level to the top metal level, to perform high frequency measurements using coplanar microwave probes. There are three contacts on each microwave probe in ground-signal-ground (GSG) configurations [9]. The center contact is the signal contact and the other two are the ground contacts with a 125 μm pitch. The configuration of the landing pads for the microwave probes is shown in Fig. 5.4(b). The landing pads are squares with 100 μm sides. A minimum of 250 μm spacing is used between two microwave probes.

Table 5.1. **Physical dimensions and number of vias present in the capacitors of the test vehicle. Nominal capacitances and via inductances are also listed.**

Capacitors	Edge length (μm)	Area of the top electrode (x 10^4 μm^2)	# of vias for each capacitor	Nominal C (pF) with Ta$_2$O$_5$ (0.38 μm)	Total via inductance* (pH)
C1	275	6	4	35	-**
C2	440	11	12	64	1.36
C3	662	28	14	163	1.1
C4	1002	57	22	332	0.87
C5	1462	113	34	658	0.48
C6	2354	283	58	1100	0.31

* Calculated in section 5.5.2. It is assumed that the total inductance is contributed by vias only.
** Could not be determined as first resonance frequency does not occur within the range (calibrated) of the impedance analyzer, 20 GHz.

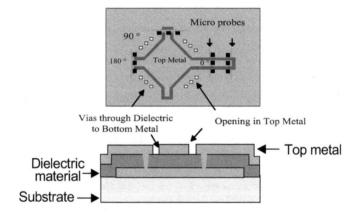

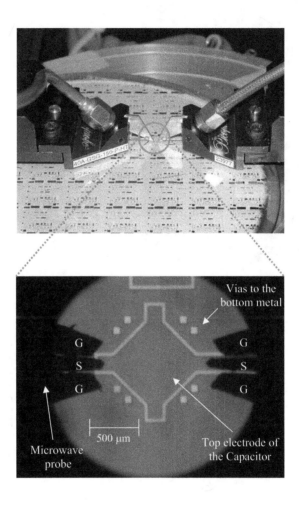

Figure 5.3. Photographs showing the arrangement of the coplanar microwave probes on the capacitor structure during S parameter measurements.

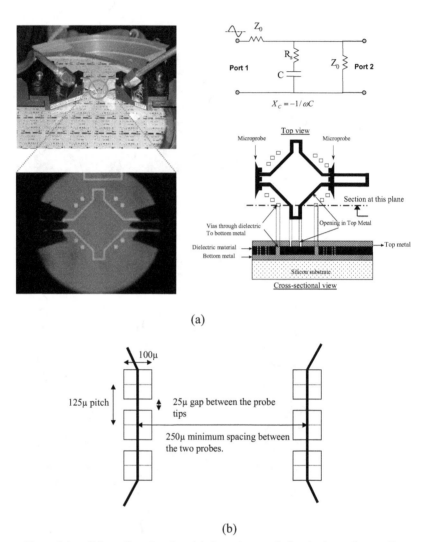

(a)

(b)

Figure 5.4. Schematics showing (a) top view and front view of capacitor structure present in the test vehicle, along with the equivalent circuit model of the capacitor (refer Fig. 5.12) for detailed equivalent circuit representation), (b) microprobe configuration and dimensions.

5.3.2 Test Vehicle Fabrication

A two mask process was used to fabricate the test structures. The substrates used in this work were 5" n-type (100) silicon wafers with resistivities in the range of 15-25 Ω-cm. The wafers were RCA cleaned. Platinum, with a target thickness of 1 μm, was deposited across the entire wafer, using electron-beam (e-beam) evaporation, to serve as the bottom electrode. A ~0.05 μm thick titanium film was used as an adhesion promoter between Si and Pt. Next, the dielectric film was deposited across the entire wafer. Different test vehicles (one for each dielectric) were fabricated for each dielectric; SiO_2, Si_3N_4, and Ta_2O_5. SiO_2 and Si_3N_4 films were deposited by plasma enhanced chemical vapor deposition (PECVD) process using a Plasmatherm reactor [10], while Ta_2O_5 films were deposited by pulsed dc magnetron reactive sputtering [11]. The process parameters used in SiO_2, Si_3N_4, and Ta_2O_5 depositions are summarized in Table 5.2.

Table 5.2.   Deposition conditions for the dielectrics used in the test vehicles

Material	Deposition process	Processing conditions
SiO_2	PECVD	SiH_4 = 400 sccm, N_2O = 900 sccm, RF power = 30 W, Pressure = 900 mTorr, Deposition temperature ~ 300 °C, Deposition rate ~ 20 nm/ min
Si_3N_4	PECVD	SiH_4 = 500 sccm, N_2= 900 sccm, He = 2250 sccm RF power = 30 W, Pressure = 900 mTorr, Deposition temperature ~ 250 °C, Deposition rate ~ 18 nm/min
Ta_2O_5	Pulsed dc reactive sputtering	O_2 = 8 sccm, Ar = 17 sccm Power = 2.5 kW, Pressure = 6.0 mTorr Deposition temperature < 100°C, Deposition rate ~ 45 nm/min

The thickness of SiO_2, Si_3N_4, and Ta_2O_5 films were around 0.31, 0.61, and 0.38 μm, respectively. Film thicknesses were measured using a variable-angle spectroscopic ellipsometer [12] and also verified by step-height measurements using a profilometer (Alpha Stepper) [13]. The dielectric films were then patterned using a standard optical lithography process [14]; the first level mask defined the via-patterns. The vias in SiO_2, Si_3N_4, and Ta_2O_5 were etched using reactive ion etching (RIE) in CHF_3/O_2 plasma at 30 mtorr pressure and 165 W plasma power. The flow rates of CHF_3 and O_2 were 45 sccm and 5 sccm, respectively. The etch rates for SiO_2, Si_3N_4, and Ta_2O_5 at these conditions were 30, 50, and 15 nm/min, respectively. Finally, blanket layers of aluminum, with a target thickness of 1 μm, were deposited using e-beam evaporation on all the test vehicles. The aluminum layers were patterned using the second mask, and then wet etched to define the top electrodes of the capacitors. The minimum feature size for etching was around 25 μm. For aluminum etching, a solution of phosphoric acid, nitric acid, acetic acid, and water in 4:2:2:1 ratio was used at 60 °C. The aluminum etch rate was approximately 100 nm/min.

5.4   Characterization Methodology

The low frequency measurements, from 1 kHz to 1 MHz, were performed using a HP4192 Impedance Analyzer [15] with 30 mV rms signal level. Standard pin probes were used for capacitance and dielectric loss measurements. One probe was contacted to the top electrode of the capacitor and other probe to bottom electrode contact. Dielectric constants were calculated using capacitance measurements, dielectric film thicknesses and the top electrode areas.

High frequency measurements, from 45 MHz to 20 GHz, were performed using a HP8510C Vector Network Analyzer with the 8341N Synthesized Sweeper and a HP8516A S-parameter set [16]. The calibration of the network analyzer with the cables, connectors, and microwave probes was made using CS5 calibration wafer. Open, short and through structures were used from the CS5 calibration wafer for the full 2-port calibration. The calibration was done over the frequency range of 45 MHz to 20 GHz, in 201 linear frequency increments. After calibration, the data were stored and used for subsequent 2-port S-parameter measurements.

In S-parameter measurements, a signal is sent across the capacitor structure from one of the microprobes (using a 50 Ω transmission line). The transmitted signal is measured at the other port (as shown in Fig. 5.4). The ratio of the magnitude of the signal that transmits through to the magnitude of the signal fed in, describes the high frequency performance of a dielectric material used in the capacitor under test. The smaller this ratio, the better the decoupling provided by the capacitor. The S_{21} measurements are transformed into insertion loss measurements using:

$$\text{Insertion loss in dB} = 20 \log (|S_{21}|) \qquad (5.4)$$

The current-voltage (I-V) characteristics of SiO_2, Si_3N_4, and Ta_2O_5 films were measured on the metal-insulator-metal (MIM) structures using an HP 4140B I-V meter [17]. Standard pin probes were used to make electrical connections to both top and bottom electrodes of the capacitor structures. Voltage was increased in steps, with a step value of 1V, until breakdown of the films occurred. A step delay of 0.5 second was maintained at each voltage step.

5.5   Results and Discussion

5.5.1   Dielectric Constant and Dielectric Loss

By convention, the complex dielectric constant of a material is

$$\varepsilon_r = \varepsilon_r^{'} - j\varepsilon_r^{''} \qquad (5.5)$$

where $\varepsilon_r^{'}$ is the real part and $\varepsilon_r^{''}$ is the imaginary part of the complex dielectric constant. The real part denotes the relative permittivity of the dielectric material, which is used in

116

calculating the capacitance. The real part, ε_r', represents the extent to which the molecules of the dielectric material can polarize when an electric field is applied, and hence determines the amount of energy that can be stored per unit volume of the dielectric material. Different polarization mechanisms, e.g., electronic, molecular, ionic, oreintational etc., contribute to the overall polarization and the dielectric constant. Each of these polarization mechanisms is frequency dependent. When an ac field is applied to a dielectric material, each polarization mechanism can effectively contribute to the net polarization up to a limiting frequency. Above that frequency, where the polarization mechanism drops out, the effective dielectric constant goes down causing dielectric dispersion.

The imaginary part, ε_r'', represents the energy lost (as heat) in the dielectric medium as the molecules polarize one way and then the other way, following the polarity of the ac field. The relative magnitude of the ε_r'' with respect to ε_r' is defined through a quantity, tanδ, called the loss tangent (or loss factor). Engineering applications of dielectrics in capacitors desire maximum ε_r' and minimum dielectric dispersion and loss tangent over the entire range of operating frequencies.

In this section, the variations in dielectric constants and loss tangents of SiO_2, Si_3N_4 and Ta_2O_5 thin films are discussed as functions of operating frequency (1 kHz to 20 GHz). In this chapter, the term 'low frequency characterization' is used for the electrical measurements from 1kHz to 1MHz, while the electrical measurements above 1 MHz are referred as 'high frequency characterization'. DC bias is used to determine breakdown field and leakage current density.

5.5.1.1 Low frequency characterization

The dielectric constants of the SiO_2, Si_3N_4 and Ta_2O_5 are plotted as functions of frequency from 1 kHz to 1 MHz in Fig. 5.5(a). For each material, the experimental results are presented for the smallest capacitors (C1). The dielectric constants are calculated using capacitance measurements, dielectric film thicknesses and the areas of the top electrodes. The experimentally determined dielectric constants of SiO_2, Si_3N_4 and Ta_2O_5 ($\sim$ 3.9, 6.7 and 23 respectively) are in close agreement with the values reported in the literature [18]. None of the dielectric materials (SiO_2, Si_3N_4 and Ta_2O_5) show significant dielectric dispersion in the low frequency range (1 kHz to 1MHz), i.e., their dielectric constants do not decrease with increase in operating frequencies. The loss tangents of SiO_2, Si_3N_4 and Ta_2O_5 films are also plotted as functions of frequency in Fig. 5.5(b). The loss tangents of all three materials are less than 0.015. A small apparent increase in dielectric constants and loss tangents for all materials near 1 MHz is due to the presence of parasitic inductances associated with the probe pins, connection leads, and test structures.

5.5.1.2 High frequency characterization

Both real and imaginary components of the transmission scattering parameters (S_{21}) are obtained from full 2-Port measurements using network analyzer. In S_{21} measurements, a signal is sent across the capacitor from one of the microprobes (using 50 Ω transmission line) and the transmitted signal is measured at the other port. Ideally, a perfect capacitor behaves as a very low impedance path at high frequency ac signals and no signal should transmit to the measuring port. The ratio of the magnitude of the signal that comes out to the magnitude of the signal fed in, describes the high frequency performance of a dielectric material used in the capacitor under test. The smaller this smaller this ratio, the better the decoupling provided by the capacitor. The S_{21} measurements are transformed into insertion loss measurements using Eq. (5.4)

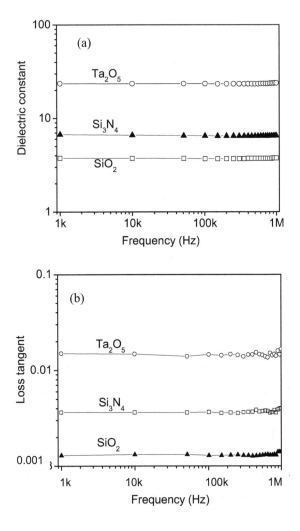

Figure 5.5. Variations of (a) apparent (measured) dielectric constant and (b) measured loss tangent of SiO₂, Si₃N₄, and Ta₂O₅ thin films, with signal frequency in low frequency range (1 kHz – 1MHz).

The insertion losses for SiO_2, Si_3N_4 and Ta_2O_5 are plotted as functions of signal frequency, varying from 45 MHz to 20 GHz, in Fig. 5.6(a), 5.6(b), and 5.6(c), respectively. For each material, the measurements are shown for six different areas of the capacitors (C1 to C6). As shown in Fig. 5.6, insertion losses for all three materials decreases with signal frequency.

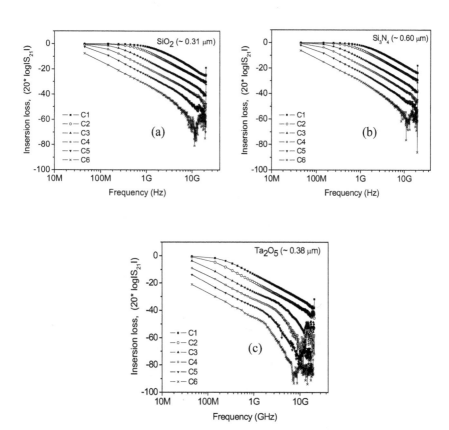

Figure. 5.6.   Insertion losses (in dB) as functions of signal frequency for SiO_2, Si_3N_4 and Ta_2O_5 thin films

For all three materials, the insertion losses at 1 GHz are also plotted as functions of capacitance on a log-log plot in Fig. 5.7. The experimental data points follow the line with a slope of −20 on a log-log and hence

$$I.L.(dB) \propto 20\log\left(\frac{1}{\omega C}\right) \propto 20\log\left[\left(\frac{d}{\omega A \varepsilon_0}\right)\left(\frac{1}{\varepsilon_r}\right)\right] \qquad (5.6)$$

where, ω is the angular frequency, A is the area of the capacitor, d is the dielectric film thickness, and ε_r is the relative permittivity of the dielectric. Thus, the decoupling provided by the same area and same thickness capacitors, is proportional to the dielectric constants of the thin film dielectrics (smaller the I.L, the better the decoupling provided by the capacitor).

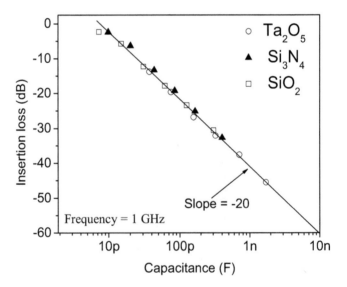

Figure 5.7. Insertion losses (in dB) at 1 GHz as functions of capacitances of SiO$_2$, Si$_3$N$_4$ and Ta$_2$O$_5$ based thin film capacitors.

121

The insertion loss drops down sharply and attains a minimum at a frequency when the inductive impedance $(X_L = \omega L)$ cancels the capacitive impedance $(X_C = -1/\omega C)$. Based on a lumped parameter model this frequency is known as the series resonance frequency and is [2]

$$f_R = \frac{1}{2\pi\sqrt{LC}} \tag{5.8a}$$

where, $\quad f_R$ = series resonance frequency

$\quad\quad\quad\quad L$ = inductance of the capacitor structure

$\quad\quad\quad\quad C$ = capacitance of the parallel plate capacitor

The resonance frequency can also be estimated from the physical dimensions of the capacitor and the effective permittivity of the dielectric material. The frequencies at which the resonance occurs are also called self-resonance frequencies. The self-resonance frequencies depend on the physical dimension of the capacitors. The first self-resonance occurs when the edge length of the capacitor is around 1/4th of the signal wavelength. The first self-resonance frequency for parallel plate capacitors is [19]

$$f_R = \frac{c}{4l\sqrt{\varepsilon_r}} \tag{5.8b}$$

where $\quad c$ = speed of light in vacuum

$\quad\quad\quad\quad l$ = dimension (edge) of the capacitor structure

$\quad\quad\quad\quad \varepsilon_r$ = dielectric constant (relative permittivity) of the dielectric

The values of first self-resonance frequencies for SiO_2, Si_3N_4 and Ta_2O_5 are calculated using Eq. 5.8(b) and plotted in Fig. 5.8. For these calculations, the dielectric constants of SiO_2, Si_3N_4 and Ta_2O_5 are taken as 3.9, 6.9 and 25 [18]. For the comparison with the calculated first self-resonance frequencies, the experimentally determined resonance frequencies from insertion loss measurements are also plotted for different size capacitors in Fig. 5.8. The experimentally determined resonance frequencies are close to the calculated first self-resonance frequencies. The experimental values of resonance frequencies for smaller capacitors, i.e., C1 to C4 for SiO_2 and Si_3N_4, and C1 for Ta_2O_5, were above the upper limit of the calibrated frequency range of the network analyzer (20 GHz) and hence could not be determined.

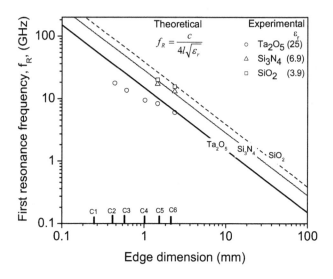

Figure 5.8. Variations of first self-resonance frequencies of SiO_2, Si_3N_4 and Ta_2O_5 based thin film capacitors as functions of capacitor edge dimension. Straight lines show the theoretical values of first self-resonance frequencies as calculated from Eq. 5(b). Data points ($\square$, Δ, O) show the experimentally determined series resonance frequencies from the insertion loss measurements.

The capacitances and hence the dielectric constants in the high frequency range (45 MHz to 20 GHz) are calculated from the insertion loss measurements, using two different circuit equivalent models (series and parallel configurations). The schematics of series and parallel circuit equivalent models used are shown in Fig. 5.9(a) and 5.9(b). Although the series and parallel equivalent models depict oversimplified representations of the actual capacitor test structures, these models provide good first order estimates of the dielectric behavior of the thin film dielectric materials.

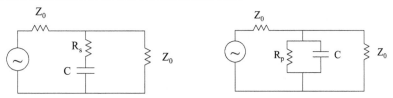

Figure 5.9.   Simplified a) series and b) parallel equivalent circuit representation (models) of the capacitors, used to extract capacitance from S-parameters.

For the series model, the capacitance value is extracted using [19]

$$C_s = \frac{-2}{Z_0\omega}\left(\frac{(1-S_{21r})^2 + S_{21i}^{\;2}}{S_{21i}}\right) \tag{5.9a}$$

and for the parallel circuit equivalent model, the apparent capacitance is extracted using

$$C_p = \frac{-2}{Z_0\omega}\left(\frac{S_{21i}}{S_{21r}^{\;2} + S_{21i}^{\;2}}\right) \tag{5.9b}$$

where Z_0 is the characteristic impedance ($50\,\Omega$), ω is the angular frequency of the applied signal, and S_{21r} and S_{21i} are the real and imaginary parts of the S_{21} scattering parameters. The procedure to extract capacitances from the S-parameter measurements is given in appendix B.

The dielectric constants for SiO_2, Si_3N_4 and Ta_2O_5 are then obtained from the capacitance measurements using film thicknesses and capacitor areas. The dielectric constants for the smallest capacitors (C1) are plotted as functions of signal frequency in Fig. 5.10(a) and 5.10(b), for series and parallel circuit equivalent models, respectively. For the small capacitors, e.g., C1 in Fig. 5.10(a) and 5.10(b), both parallel and series models show that the dielectric constants are nearly invariant over the whole range of signal frequencies (45 MHz to 20 GHz). However, small apparent variations in dielectric constants are observed at high frequencies in the 15-20 GHz range. These apparent variations in dielectric constants at high frequencies are caused by the parasitic components in the high-frequency test structures and also in the testing apparatus. These parasitic components have been ignored in the simplified series and parallel circuit equivalent representations of the test structures.

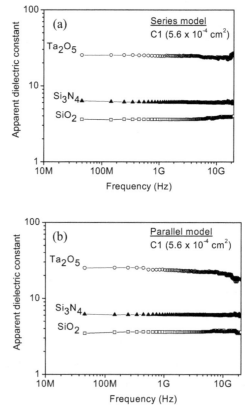

Figure 5.10. Apparent (or measured) dielectric constants as functions of signal frequency (45 MHz to 20 GHz) from the S_{21} measurements for the smallest capacitor, C1, for (a) series and (b) parallel configurations in Fig 5.9.

The dielectric constants for the largest capacitors (C6) are also plotted as functions of signal frequency in Fig. 5.11(a) and 5.11(b), for series and parallel circuit equivalent models respectively.

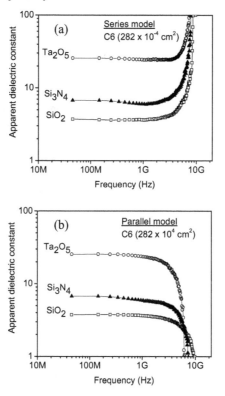

Figure 5.11. **Apparent (measured) dielectric constants as functions of signal frequency (45 MHz to 20 GHz) from the S_{21} measurements for the largest capacitor, C6, for (a) series and (b) parallel configurations in Fig 5.9.**

Although little or no dispersion in the dielectric constant is observed for small size capacitors, apparent variations in the dielectric constants are evident for large size capacitors at higher frequencies. For series equivalent model, the apparent dielectric constants of all three materials unrealistically increase with signal frequency beyond 3 GHz for C6 capacitors. On the other hand, an apparent decrease in dielectric constants of all three materials is evident from parallel equivalent model. These apparent variations in dielectric constants are not real and do not correspond to the dielectric dispersions of the materials because no dispersion was observed for small size capacitors in Fig. 5.10(a) and 5.10(b).

The large size capacitors, e.g., C6, exhibit first resonance frequency at around 3 GHz. Hence, around and beyond 3 GHz, the impedances due to capacitive and inductive (parasitic) components are comparable. Thus, estimation of dielectric constants based on oversimplified series and parallel circuit equivalent models (with no inductive components) give unrealistic values near and beyond first-resonance frequencies. A more complete model, as shown in Fig. 5.12 having combinations of both series and parallel models along with an inductor in series and allowance for the Z_x for the representation of distributed parameters, resonance modes etc., might give realistic estimates of dielectric constants even at high frequencies. However, because of the number of unknowns involved, the determined values of these unknowns using data obtained on the present test vehicle would be unreliable.

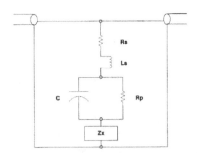

Figure 5.12.   Equivalent circuit representation of the parallel plate capacitor, having a combination of both series and parallel circuit configurations along with series inductance. Z_x represents the allowance for the representation of distributed parameters, resonance modes etc.

In summary, no or insignificant dispersion in the dielectric constants of SiO_2, Si_3N_4 and Ta_2O_5 is indicated. Hence, the polarization mechanisms responsible for the dielectric behavior of these materials can follow the applied alternating electric fields at least up to 20 GHz. Among various polarization mechanisms, combination of electronic and molecular (or atomic) polarization mechanisms appears to be the dominant mechanisms responsible for the dielectric constants of SiO_2, Si_3N_4 and Ta_2O_5. The restoring force in electronic and molecular (or atomic) polarizations is directly proportional to the separation distance of the induced dipoles, and thus these polarization mechanisms can effectively respond to the alternating electric fields at least until tera-hertz ($> 10^{12}$ Hz) range [20]. In case of the polarization due to the orientation of dipoles, there is no direct mechanical force that 'pulls' the dipoles back to the random orientation, and hence contribution due to orientation polarization dies out at several hundred MHz frequencies [21]. Nearly no variation in dielectric constant over wide range of frequency (1kHz to 20 GHz) confirms that the contribution of orientation dipoles towards the dielectric constants of SiO_2, Si_3N_4 and Ta_2O_5 is minimal.

5.5.2 Total Series Inductance

In the given test vehicle, several vias are provided in parallel to bring up the electrical connections from the bottom metal level to the top metal level, for high frequency measurements using coplanar microwave probes. The number of via connections provided in different sized capacitors is listed in Table 5.1. These metal filled vias act as inductor connections in series to the rest of the parallel plate capacitor structures. Apart from the via-inductances, the rest of the structure also contributes to the total series inductances. In this thesis, it is assumed that most of the total measured inductance is contributed by the vias.

The total series inductances (due to via connections and rest of the structure, L) associated with the capacitor structures are estimated from the first self-resonance frequencies, using Eq. 5.8(a). Total inductances are plotted against the number of vias for Ta$_2$O$_5$ capacitors (C2 to C6) in Fig. 5.13. The total inductance decreases linearly with the number of vias. For comparison, a line with slope of 1 is also plotted along with the experimental data points in Fig. 5.13. The close agreement between experimental data points with the line of slope 1 confirms that the parallel vias connections indeed act as parallel inductors and the total inductance is inversely proportional to the number of vias present in the capacitor structures. Hence, providing several via connections in parallel can reduce the total inductance of the capacitor structures. The via-inductances are also normalized per micron of via-length and plotted in Fig. 5.13. The extrapolated results show that the via-inductance is around 45 pH/via per μm of via-length.

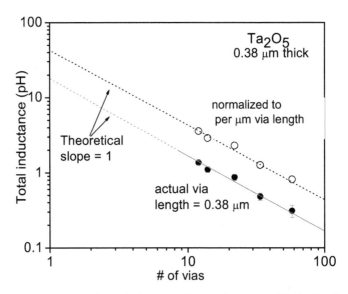

Figure 5.13. Variation of total inductance with number of parallel vias for Ta$_2$O$_5$ based capacitors. The normalized via inductances per μm of via length are also plotted.

The total inductance per μm is plotted as a function of aspect ratio of the vias in Fig. 5.14. The via-lengths for SiO_2, Si_3N_4 and Ta_2O_5 capacitors are 0.31, 0.60 and 0.38 μm respectively and the via-edges are 50 μm. The total inductances are around 45-50 pH/μm for all SiO_2, Si_3N_4 and Ta_2O_5 based capacitors. The inductances of the vias are also estimated, to a first order approximation, assuming them to be cylindrical metal lines surrounded by ground planes on both sides [22]. The via-inductances estimated using these models are about an order and a half magnitude lower than the total inductances estimated from our test structures. Goldfarb et al. [23] and Wu et al. [24] have also shown sub-pH/μm via-inductances can be achieved with cylindrical vias surrounded by the ground planes.

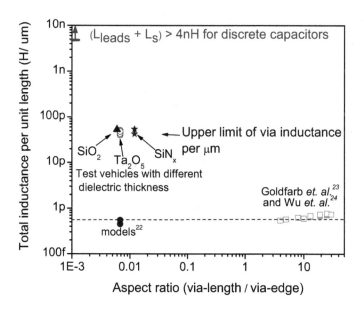

Figure 5.14.   Inductance per μm of via length as functions of via aspect ratio

In our test vehicles, the total inductances of the whole test structures are reported, which is partially due to via-inductance and partially due to the rest of the structure configurations and test probes. Hence, apparently higher estimates of the total inductances compared to the calculated via-inductances can be considered as the 'upper limit' of the via-inductances for our test structures. Nonetheless, the total inductances associated with the embedded thin film capacitors are about 2 to 3 orders of magnitude lower than most state-of-the-art surface mount discrete chip capacitors [8]. Although the self-inductances of the discrete capacitors are about 60-100 pH, the leads connecting the discrete capacitors to the IC add more than 4 nH of lead inductance. These lead connections are avoided by using thin film embedded capacitors right beneath the ICs. Hence, overall inductance can be reduced from several nH to pH range by using thin film embedded capacitors near the ICs.

5.5.3 Leakage Current Density

The dc leakage current densities for SiO_2, Si_3N_4 and Ta_2O_5 are plotted as functions of applied electric fields in Fig. 5.15. The leakage current densities for all three materials can be broadly classified into three main regimes; ohmic, nonlinear and breakdown regime.

The ohmic region exists at low electric fields (< 0.3 MV/cm) with linear variation of leakage current density with applied electric field. The leakage current densities for all three materials are below 10^{-7} A/cm^2 in the ohmic regime. This ohmic conduction is attributed to thermal excitation or hopping of trapped electrons from one trapped site to another trapped site. This region is followed by a nonlinear conduction mechanism regime (> 0.3 MV/cm). The conduction mechanism in the nonlinear conduction regime is speculated to be a combination of various bulk and electrode limited conduction mechanisms; e.g., Schottky barrier limited mechanism, Poole-Frenkel emission and space charge limited conduction mechanisms [25].

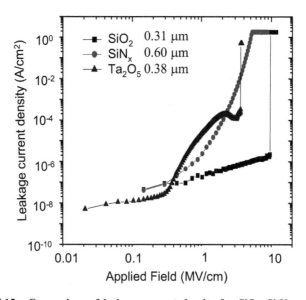

Figure 5.15. Comparison of leakage current density for SiO₂, Si₃N₄ and Ta₂O₅ thin films as a function

Leakage current in both ohmic and nonlinear regions is reversible with respect to applied electric field. In the non-linear regime, SiO_2 shows lower leakage current density compared to Si_3N_4 and Ta_2O_5. At still higher fields, the leakage current increases sharply with applied field until the film breaks down irreversibly. Films after breakdown are permanently electrically shorted and show a leakage current above the current compliance limit (10^{-2} A) even at very low voltages.

5.5.4 Capacitance Density and Breakdown Field

Besides high capacitance densities, dielectric thin films should also exhibit good insulating behavior or low dc conductivity, so that the stored charges are not simply conducted from one plate of the capacitor to the other through the dielectric. Hence, lower leakage current densities higher breakdown fields of dielectric films are desirable.

The capacitances of SiO_2, Si_3N_4 and Ta_2O_5 based capacitors at 1 MHz are plotted as functions of the capacitor areas in Fig. 5.16. The capacitances increase linearly (from pF to nF range) with capacitor areas over two orders of magnitude, for all three dielectric materials (SiO_2, Si_3N_4 and Ta_2O_5). Hence a wide range of capacitances (pF to nF range) can be achieved using thin film capacitors.

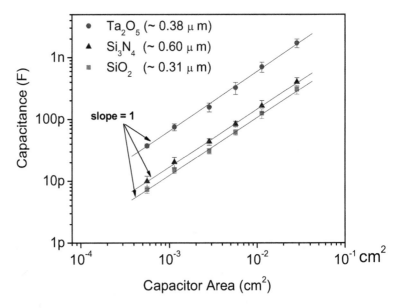

Figure 5.16. Variations of measured capacitances at 1 MHz as functions of capacitor areas for SiO_2, Si_3N_4 and Ta_2O_5 based capacitors. Dielectric thicknesses are shown in parenthesis along with the dielectric material label.

The electrical breakdown fields and dielectric constants of SiO_2, Si_3N_4 and Ta_2O_5 thin films are shown in Fig. 5.17. A trade-off between breakdown field strengths and dielectric constants of SiO_2, Si_3N_4 and Ta_2O_5 thin films exist. The breakdown fields of all three materials closely follow the 'best can achieve' line (derived and explained in detail in chapter 4) [26,27].

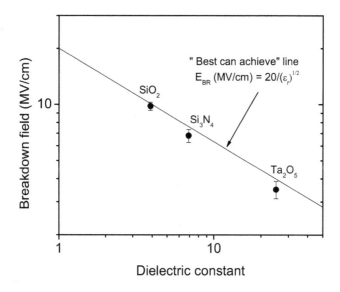

Figure 5.17. Breakdown field as a function of dielectric constant. Breakdown field decreases as dielectric constant increases.

5.6 Summary and Conclusions

In this chapter, the high frequency performance (up to 20 GHz) and dc characteristics of three dielectric materials (SiO_2, Si_3N_4, and Ta_2O_5) are discussed for thin film decoupling capacitors. All three selected dielectric materials for this study are compatible with current silicon processing technology. From impedance and insertion loss measurements, the dispersion in dielectric constants of all three materials is evaluated from 1 kHz to 20 GHz. The total inductance of the power connect vias is estimated from first self-resonance frequencies. The trade-off between capacitance density and breakdown voltage is discussed for thin film dielectrics. The dc leakage current densities of SiO_2, Si_3N_4, and Ta_2O_5 are also compared.

The dielectric constants of SiO_2, Si_3N_4, and Ta_2O_5 exhibit no dispersion up to at least 20 GHz of signal frequency. No dispersions in dielectric constants until 20 GHz indicate that electronic and molecular (atomic) polarization mechanisms are responsible for the dielectric response of these dielectrics. The dielectric losses for all three materials are less than 0.015.

The total inductance of power connect vias is determined to be less than 50 pH/μm of via, which is at least two orders of magnitude lower than most discrete capacitors with lead connections (> 4 nH). Several vias in parallel further reduce the overall inductance of the thin film capacitors. In addition to low self-inductance, the shorter connections and elimination of connection leads in thin film decoupling capacitors make them highly suitable for integration into 2D and 3D structures [28]. Thin film capacitors based on SiO_2, Si_3N_4, and Ta_2O_5 can provide capacitance densities comparable to the state-of-the-art discrete capacitors. The leakage current densities below 10^{-6} A/cm^2 (at 0.5 MV/cm) can be achieved with SiO_2, Si_3N_4 and Ta_2O_5 thin film dielectrics. The breakdown fields and dielectric constants of SiO_2, Si_3N_4 and Ta_2O_5 closely follow the empirically determined "best can achieve" relationship E_{BR} (MV/cm) $= \left(20/\sqrt{\varepsilon_r}\right)$.

5.7 References

5.1   R. K. Poon, *"Computer Circuits Electrical Design,"* Prentice Hall, Eaglewood Cliffs, 1995, ch. 8, pp 195-229.

5.2   W. D. Brown, *"Advanced Electronic Packaging with Emphasis on Multichip Modules,"* IEEE Press Series on Microelectronic Systems, 1998,ch. 3, pp. 133-136.

5.3   H. B. Bakoglu, *"Circuits, Interconnections, and Packaging for VLSI,"* Addison-Wesley VLSI Systems Series, ch 7, pp. 281-335.

5.4   B. K. Gilbert and G.- W. Pan, IEEE Trans. Microwave Theory Tech. **45**, 1819 (1997).

5.5   R. R. Tummala, E. J. Rymaszewski, and A. Klopfenstein (Eds), *"Microelectronics Packaging Handbook,"* 2nd ed., vol 2, New York: Chapman and Hall, 1997, ch 3, pp. 232-270.

5.6   R. R. Tummala, *"Fundamentals of Microsystems Packaging,"* McGraw-Hill, 2001, ch 4, pp. 120-182.

5.7   M. P. Goetz, IEEE Trans. Comp., Packag., Manufact. Technol. B**19**, 518 (1996).

5.8   J. N. Humenik, J. M. Oberschmidt, L. L. Wu, and S. G. Paull, IBM Journal of Research and Development. **36**(5), 935 (1992).

5.9   Cascade Microtech's Air Coplanar™ Probes (ACP), GSG tip- 125μm pitch.

5.10   Plasmatherm-73, Plasmathrm Inc.

5.11   CVC Products Inc., Rochester, NY 14603.

5.12   VASE®, J.A. Woollam Co., Inc. VASE®

5.13   alpha-stepper® 200, Tencor Instruments, part # 035815 Rev. E., 1983 release.

5.14   6100 B DSW wafer stepper(R) system.

5.15   4192A, LF Impedance Analyzer, 5Hz-13MHz, Yokogawa-Hewlett Packard, Ltd.

5.16   HP 8510 Vector Network analyzer, Yokogawa-Hewlett Packard, Ltd.

5.17   4140 pA meter / DC voltage and current source, Yokogawa-Hewlett Packard, Ltd.

5.18   L. I. Maissel, and R. Glang, *"Handbook of Thin Film Technology,"* 1st ed. (reissue), McGRAW-HILL book Co., 1983, ch.16, pp. 21-33.

5.19   Peter A. Rizzi, *Microwave Engineering,* (Prentice-Hall 1988),ch. 1-3, pp 1-102.

5.20   D. D. Pollock, "Physical Properties of Materials for Engineers," 2nd Edition, CRC Press, 1993, ch. 12, pp. 554-571.

5.21   W. D. Callister, Jr., *"Materials Science and Engineering – an Introduction,"* 3rd Edition, John Willy & Son, Inc. 1994, ch.19, pp. 628-632.

5.22   S. Walker, " Capacitance, inductance and crosstalk analysis," 1990, Chapter 2, pp 85-121.

5.23   M. E. Goldfarb and R. A. Pucel, "Modeling via hole grounds in microstrip," IEEE microwave and guided wave letts., vol. 1, no. 6, 1991.

5.24   J. H. Wu and J. A. del Alamo, "A high aspect-ratio silicon substrate-via technology and applications," MARCO Review, March 22,2002.

5.25   D. R. Lamb and D. Robert, *"Electrical conduction mechanisms in thin insulating films,"* (Methuen, London 1967), Chap. 1-6.

5.26   P. Jain and E. J. Rymaszewski, IEEE Trans. Adv. Packag., **25**, 454 (2002).

5.27   P. Jain, J. Y. Kim, Y. Xiao, R. Natarajan, E. J. Rymaszewski, R. J. Gutmann, and T. P. Chow, *Proc. CPES annual review* 2000, Blacksburg, VA, pp. 155-158, 2000.

5.28   J.-Q. Lu, Y. Kwon, A. Jindal, K.-W. Lee, J. McMahon, G. Rajagopalan, A.Y. Zeng, R. P. Kraft, B. Altemus, B. Xu, E. Eisenbraun, J. Castracane, J. F. McDonald, T. S. Cale, A. Kaloyeros, and R.J. Gutmann, (invited paper), at 19th International VLSI Multilevel Interconnection (VMIC) Conference, pp.445-454,Singapore, November 18 - 20, 2002.

CHAPTER 6

ELECTRODE METAL DIFFUSION IN TANTALUM OXIDE

On-chip power distribution noise can be suppressed and controlled effectively by integrating (or embedding) thin film decoupling capacitors in the multi-level metallization (MLM) of the ICs [1-3]. Using embedded capacitors at on-chip level, the distance between decoupling capacitors and switching transistors can be brought down to few hundred nanometers. The close proximity of decoupling capacitors to the switching transistors offers a reduction in parasitic inductances to pH range, which directly transforms into 2 to 3 orders of magnitude reduction in power distribution noise, especially at GHz switching frequencies.

A cost-effective solution for the on-chip integration of decoupling capacitors demands minimum alteration in the already defined/established processing steps and design rules for BEOL process flow. Only selected metals, such as Cu, Al, Ti, and Ta, can be used in the MLM layers due to their compatibility with the rest of the BEOL processing steps [4]. Thus, the choice of electrode metals for embedded decoupling capacitors is limited to these selected metals or their metallic compounds, e.g., TaN, TiN etc. Choosing capacitor electrodes to be the same metal used for on-chip interconnects facilitates the on-chip capacitor integration, due to reduced integration risks. Currently, copper (Cu) is replacing aluminum (Al) as the choice of on-chip interconnects metallization, due of its lower resistivity and superior resistance to electromigration [4-6]. Hence, successful on-chip integration of thin film capacitors demands the compatibility of the high-κ dielectrics with Cu.

The major reliability problem with Cu is its fast drift and diffusion into low-κ interlayer dielectrics (ILDs), such as SiO_2 [7-9], SiN_x [10,11], polymers [12-14] etc. Cu diffusion through dielectrics causes degradation of electrical properties, such as leakage current density, breakdown field etc. Tantalum (Ta) and Ta based compounds, such as TaN, TaO_xN_y etc., are used as barrier layers to contain Cu-diffusion into dielectrics [15, 16]. Similarly, Cu-diffusion into the dielectrics of embedded capacitors leads to degradation in electrical properties, such as leakage current density, breakdown field, dielectric loss, etc., with time. The presence of high electric field due to dc bias across the capacitor further aggravates electrode metal (Cu) diffusion into the dielectric. Hence, in order to use tantalum oxide (Ta_2O_5) for on-chip decoupling capacitors, Cu-diffusion into it must be contained.

6.1 Overview of the Chapter

In this chapter, the extent of Cu diffusion/drift into tantalum oxide films is determined and compared with Al, Ta, and Ti at various biasing and temperature conditions using two electrical testing techniques: bias-temperature-stress (BTS) and triangular voltage sweep (TVS). It is important to recognize that both metal ion and neutral atoms may penetrate into the dielectric. Electrical bias-enhanced diffusion of metal ions is sometimes also referred as "drift" in the literature. Nevertheless, two terms "diffusion" and "drift" are synonymously used in the literature and also in this thesis. These techniques are discussed in detail in section 6.2. The extent of Cu-diffusion and measures to contain Cu-diffusion are discussed in section 6.3. The key results and conclusions are summarized in section 6.4.

6.2 Experimental Methodology Used

6.2.1 Metal-Insulator-Semiconductor (MIS) Capacitors: Structure and Fabrication

The schematic of the metal-insulator-semiconductor (MIS) capacitors used to determine metal diffusion into the dielectric is shown in Fig. 6.1. First, a thin layer of SiO_2 (~ 25 nm) was thermally grown on a p-type silicon wafer. It is followed by the deposition of Ta_2O_5 (~ 400 nm) using pulsed dc reactive sputtering at 5 kW sputtering power and 6 mtorr pressure. A thin layer of SiO_2 between Ta_2O_5 and p-doped silicon provides a clean oxide-semiconductor interface, free from interface traps. Metal (Cu, Al, Ti, and Ta) was sputter or e-bream deposited on the top of Ta_2O_5 using a shadow mask containing circular apertures 0.5 mm in diameter. The backside of the silicon substrate was puddle-etched in buffered oxide etch (BOE) to remove residual oxide and lined with a ~300 nm Al layer to obtain ohmic contact with the grounded chuck. MIS capacitor structures were annealed in Ar/H_2 ambient to get rid of any interface and trapped charges.

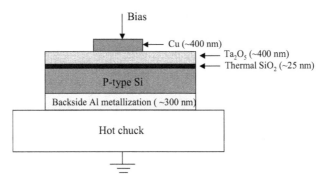

Figure 6.1. **Schematic of MIS capacitors used in BTS and TVS testing to determine electrode metal (Cu) diffusion into the dielectric.**

6.2.2 Capacitance-Voltage (C-V) Characteristics of MIS Capacitors

The electrical behavior of the generic MIS capacitor is described below. For the sake of discussion the substrate is assumed to be p-type. When swept from negative to positive voltage, the capacitor goes from (a) accumulation to (b) depletion to (c) inversion [17]. The charge distributions at the insulator-oxide interface in different regions are shown in Fig. 6.2(a). The total capacitance of the MIS capacitor depends on the insulator and semiconductor capacitances. The resulting capacitance-voltage (C-V) characteristics are shown in Fig. 6.2(b).

Flat-band voltage (V_{FB}) in the C-V curve is particularly important to determine metal diffusion/drift through the dielectric. The flat-band voltage is determined by the metal-semiconductor work function difference ϕ_{MS} and the various oxide charges (lumped together as Q_i) through Eq. 6.1. The charge effects consist of Q_f, Q_m, Q_{it}, and Q_t (fixed, mobile, interface and trapped charges in the insulator respectively), and are represented schematically in Fig. 6.3.

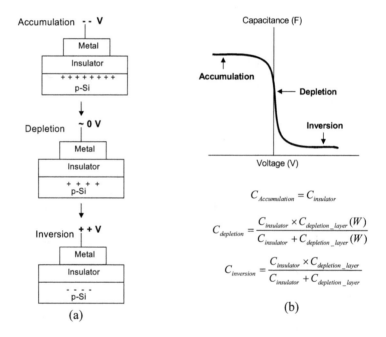

$$C_{Accumulation} = C_{insulator}$$

$$C_{depletion} = \frac{C_{insulator} \times C_{depletion\_layer}(W)}{C_{insulator} + C_{depletion\_layer}(W)}$$

$$C_{inversion} = \frac{C_{insulator} \times C_{depletion\_layer}}{C_{insulator} + C_{depletion\_layer}}$$

(a)   (b)

Figure 6.2. **(a) Accumulation, depletion and inversion regions of a MIS capacitor under the influence of voltage sweep. (b) Capacitance-Voltage**

characteristic of MIS capacitor: accumulation and inversion capacitances are constant, whereas depletion capacitance is a function of the depletion layer width "W".

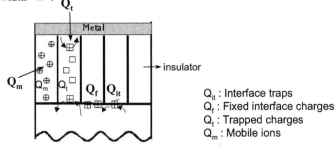

Q_{it} : Interface traps
Q_f : Fixed interface charges
Q_t : Trapped charges
Q_m : Mobile ions

Figure 6.3. **Schematic of charge effects in insulator bulk and at interfaces. Arrows represent exchange of charge at interface occurring with traps and interface states**

If the insulator capacitance is C_i, then mathematically,

$$V_{FB} = \phi_{MS} - \frac{Q_i}{C_i}$$ (6.1)

which can be expanded as

$$V_{FB} = \phi_{MS} - \frac{Q_f}{C_i} - \frac{Q_{it}}{C_i} - \gamma \frac{Q_t}{C_i} - \gamma \frac{Q_m}{C_i}$$ (6.2)

where

$$\gamma = \frac{\int_0^t \rho(x) * x dx}{t \int \rho(x) dx}$$ (6.3)

γ describes the charge distribution in the dielectric, where ρ is the charge density, x is the distance from gate interface and t is the total dielectric thickness.

6.2.3 Bias-Temperature Stress (BTS) Technique

BTS technique is used to determine mobile charges. In this technique, the MIS capacitor is heated to 150-300 °C to make the charges mobile, and a gate bias is applied for a time sufficiently long for charges to drift to the oxide interface. The MIS capacitor is then rapidly cooled to room temperature under bias and a C-V curve is measured. The procedure is also repeated with opposite bias polarity. The mobile charges are determined from the negative flat-band voltage shift, given by

$$\Delta V_{FB} = -\gamma \frac{\Delta Q_m}{C_i}$$

(6.4)

The C-V curve shifts horizontally due to the change in flat-band voltage after BTS cycle. Since only relative changes are important, the exact V_{FB} was not measured in this thesis work. Instead, voltage corresponding to 50% of maximum capacitance was used in the calculations to determine mobile ion concentration.

BTS C-V measurements were made on HP 4280A 1 MHz Capacitance Meter/C-V Plotter. A small a.c signal of 10 mV r.m.s was superposed on the applied dc bias. Capacitors were vacuum-held on an MSI Electronics Light Shield/ Hot Chuck and were under nitrogen purge throughout BTS (Fig. 6.4). The samples were biased at high temperatures, water-cooled rapidly down to room temperature (with bias on) at periodic intervals and C-V measurements were conducted. Biasing voltages ranged from 0.25 MV/cm to 0.75 MV/cm and temperatures from 50 to 300 $^{\circ}$C.

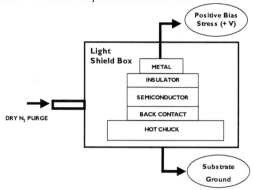

Figure 6.4. **Schematic of BTS testing setup**

6.2.4 Triangular Voltage Sweep (TVS) Technique

TVS is based on measuring charge flow through the insulator at elevated temperature in response to an applied time-varying voltage. First, a positive gate voltage is applied to a capacitor at high temperature, allowing the mobile ions (if present or if injected from gate metal) to accumulate at the opposite interface. Then a slow linear voltage ramp α (from positive to negative gate voltage) is applied. The biasing scheme is shown in Fig. 6.5. If the ramp rate is sufficiently low, then the measured current is the sum of displacement current and conduction current due to mobile charge, given by

$$I_{measured} = I_{displacement} + I_{Mobile} \tag{6.5}$$

The resulting $I_{displacement}$ (displacement current due to discharging of the capacitor) is directly proportional to the low frequency capacitance C_{low-v}.

$$I_{displacement} = \left(\frac{dQ}{dt}\right) = C_{low-v}\left(\frac{dV}{dt}\right) = C_{low-v}\alpha \tag{6.6}$$

Leakage current (dc) $I_{leakage}$ results in an extra component that distorts the $I_{measured}$ (measured current) profile. Coherent motion of the accumulated mobile ions under the changing voltage (near 0 V bias) contributes to current peaks (I_{mobile}) superposed on the low frequency C-V characteristic. The amount of drifted charge Q_{mobile} can be determined from the integrated area under the peak (Fig. 6.5). Hence,

$$Q_{mobile} = \frac{\int I_{mobile} dV}{\alpha} \tag{6.7}$$

TVS scans were performed using the HP 4140B pA Meter. The capacitors were vacuum-held on an MSI Electronics Light Shield/ Hot Chuck and were under nitrogen purge throughout the experiment. TVS voltage sweeps were made at 50-300 °C, and at 0.01 V/s. After an initial bias (equivalent to 0.75 MV/cm) was applied for a given

144

interval, voltage scans were performed at high temperature from +20 V to -30 V to detect peak features. The current (I) values were converted to capacitance (C) using the relation $I = \alpha C$, where α is the voltage sweep rate.

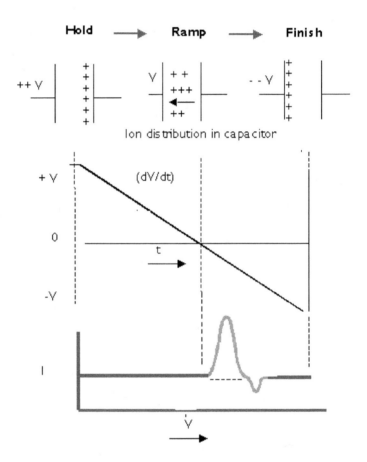

Figure 6.5. **TVS technique: peak in I-V characteristic results from mobile ion motion across the insulator**

6.3   Results and Discussion

6.3.1   Al/ SiO$_2$/ Si and Cu/ SiO$_2$/ Si Capacitors

As mentioned in section 6.2.1, a thin film of thermally grown SiO$_2$ was used between Ta$_2$O$_5$ and Si in MIS capacitors for BTS and TVS testing. This was done because only thermal SiO$_2$ forms a clean interface with Si, free from interface traps. Otherwise, the C-V curve of MIS capacitors will have a distorted shape due to poor Si-insulator interface. This makes the analysis of the results extremely difficult and prone to errors. Therefore, MIS capacitors with thermally grown SiO$_2$ (~25 nm) as the only insulator layer between metal (gate) and Si were made. Cu and Al diffusion/drift into thermally grown SiO$_2$ was tested on these capacitors using BTS and TVS techniques. Later, the results obtained from these capacitors are used to separate the contribution of SiO$_2$ from that of Ta$_2$O$_5$ in Al/Ta$_2$O$_5$/ SiO$_2$/ Si and Cu/Ta$_2$O$_5$/ SiO$_2$/ Si capacitors (in section 6.3.2 and 6.3.3). These results also provide a baseline to compare Cu diffusion in Ta$_2$O$_5$ and thermally grown SiO$_2$.

Capacitance-voltage (C-V) curves were measured before and after bias-stressing Al/SiO$_2$/Si and Cu/SiO$_2$/Si capacitors, for different durations of biasing time. The structures were tested at 2.0 MV/cm for 120 min at 150 $^{\circ}$C. The results are shown in Fig. 6.6(a) and 6.6(b). In both cases (Al/Ta$_2$O$_5$/SiO$_2$/ Si and Cu/Ta$_2$O$_5$/SiO$_2$/ Si capacitors), no negative shift of flat-band voltage was detected. Hence, no mobile ions were detected by BTS technique in the thermally grown SiO$_2$. This is consistent with the results reported in the literature [7,18,19]. As shown in Fig. 6.6(a) and 6.6(b), C-V curve shifts to the right after first BTS test cycle, but stays there during subsequent cycles. This could be due to the electron trapping at Si/dielectric interface or damage healing at metal/dielectric interface due to thermal cycle. Mobile ions were not detected with TVS test either. Hence, it is concluded that both Cu and Al don't penetrate into thermal SiO$_2$ at least till 2 MV/cm at 150 $^{\circ}$C.

146

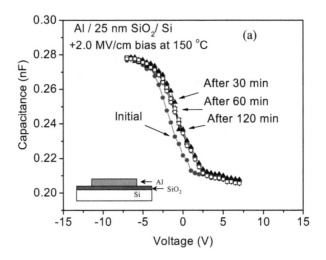

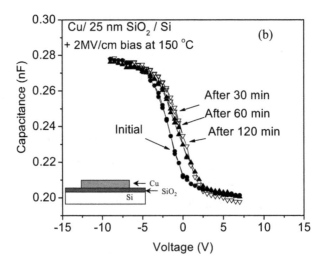

Figure 6.6. BTS results: (a) Al/SiO₂/Si and (b) Al/SiO₂/Si capacitors. No negative shift in the flat-band voltage observed after BTS cycles till 120 min at 150 °C.

6.3.2 Al/Ta$_2$O$_5$/SiO$_2$/Si and Cu/Ta$_2$O$_5$/SiO$_2$/Si Capacitors

As discussed in previous chapters, Ta$_2$O$_5$ electrically breaks down around 3 MV/cm. High dc leakage currents near breakdown field makes the capacitance measurements unreliable. Therefore, as a guideline, the bias stressing was kept below 1/3rd of the breakdown field of the dielectric. BTS and TVS tests were performed on Al/Ta$_2$O$_5$/SiO$_2$/Si and Cu/Ta$_2$O$_5$/SiO$_2$/Si capacitors at 0.25, 0.5, and 0.75 MV/cm and temperatures varying from 50 °C to 300 °C.

6.3.2.1 BTS Testing Results

(a) Al/Ta$_2$O$_5$/SiO$_2$/Si capacitors

The C-V curves for Al/Ta$_2$O$_5$/SiO$_2$/Si capacitors were measured before and after bias stressing at different intervals of time. No negative shift of flat-band voltage was detected, when tested at 0.25, 0.5 and 0.75 MV/cm for 120 min at 300 °C. The representative behavior of C-V curves corresponding to 0.75 MV/cm bias at 300 °C is shown in Fig. 6.7(a). Thus no mobile ions were detected in Ta$_2$O$_5$ by BTS testing. Like Al/SiO$_2$/Si capacitors, C-V curves for Al/Ta$_2$O$_5$/SiO$_2$/Si capacitors shifted slightly towards the right after first BTS cycle, but stayed there during subsequent BTS cycles. As mentioned earlier, the C-V shift towards the right could be due to electron trapping at Si interface. Flat-band voltage does not shift to the right after first BTS cycle, if negative bias is applied during BTS test (shown in Fig. 6.7(b)).

Hence, Al does not diffuse into Ta$_2$O$_5$ at least till 0.75 MV/cm bias applied for 120 minutes at 300 °C. The formation of dense Al$_2$O$_3$ at the Al/Ta$_2$O$_5$ interface inhibits the penetration of Al or Al ions into Ta$_2$O$_5$. Therefore, in this work, Al/Ta$_2$O$_5$/SiO$_2$/Si capacitors are used as control samples to determine Cu-diffusion into Ta$_2$O$_5$.

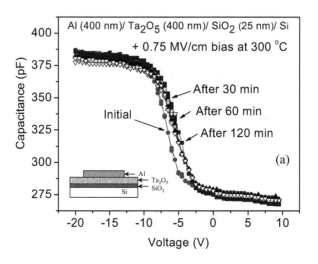

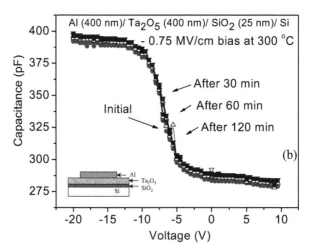

Figure 6.7. BTS results on Al/Ta$_2$O$_5$/SiO$_2$/Si capacitors at 300 °C with bias of (a) 0.75 MV/cm and (b) –0.75 MV/cm. No negative shift in the flat-band voltage observed after BTS cycles till 120 min at 300 °C.

(b)   Cu/Ta₂O₅/SiO₂/Si capacitors

The BTS results for Cu/Ta₂O₅/SiO₂/Si capacitors at 0.75 MV/cm and 150 °C are shown in Fig. 6.8. The C-V curves are measured after 0, 30, 90, and 120 min of biasing at 150 °C. No negative shift in flat-band voltage confirms the absence of Cu diffusion into Ta₂O₅ at 0.75 MV/cm and 150 °C (normal BTS testing conditions) till 120 min.

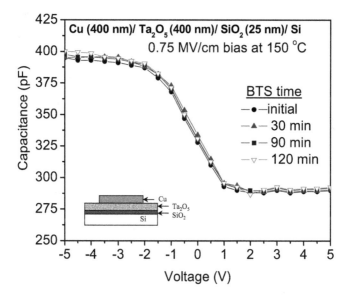

Figure 6.8.   BTS results on Cu/Ta₂O₅/SiO₂/Si capacitors at 150 °C with bias of 0.75 MV/cm. No negative shift in the flat-band voltage observed after BTS cycles till 120 min.

However, C-V curves show a negative shift in flat-band voltage, when tested at 300 °C and 0.75 MV/cm bias. The systematic negative shift in C-V curves with BTS time is shown in Fig. 6.9(a). The negative flat-band shifts at different biasing conditions (0.25, 0.5, and 0.75 MV/cm) at 300 °C are plotted as functions of BTS time in Fig. 6.9(b).

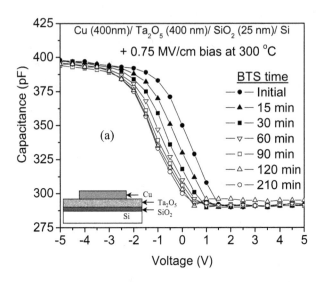

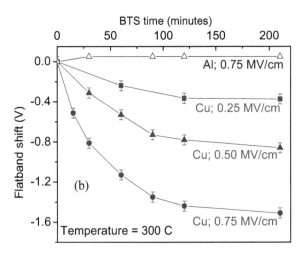

Figure 6.9. (a) BTS results on Cu/Ta₂O₅/SiO₂/Si capacitors at 300 °C with bias of 0.75 MV/cm. Flat-band voltage systematically shifts in the negative direction. (b) Flat-band shifts as a function of BTS time for different biasing conditions at 300 °C.

In order to confirm that this negative shift is due to mobile ions (Cu$^+$ or Cu^{++}), the BTS testing was performed with –0.75 MV/cm on the metal (gate). Fig. 6.11 shows that there is no shift in the flat-band voltage if the applied bias is negative.

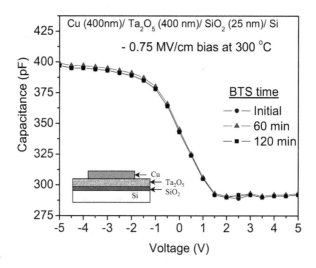

Figure 6.10. BTS results on Cu/Ta$_2$O$_5$/SiO$_2$/Si capacitors at 300 °C with - 0.75 MV/cm bias. No shift in flat-band voltage is detected.

6.3.2.2   TVS Characterization Results

In TVS characterization, first 0.75 MV/cm was applied to Al/Ta$_2$O$_5$/SiO$_2$/Si and Cu /Ta$_2$O$_5$/SiO$_2$/Si capacitors at 300 °C for different durations of time from 0 to 120 min. Voltage scans were performed at 0.1 V/s from +20 V to -30 V. The TVS characterization results (I-V curves) are shown in Fig. 6.11. The absence of any peaks in Al/Ta$_2$O$_5$/SiO$_2$/Si capacitors (Fig. 6.11(a)) reconfirms the BTS results that Al does not diffuse into Ta$_2$O$_5$ even at aggressive test conditions. Fig. 6.11(b) shows the TVS results for Cu /Ta$_2$O$_5$/SiO$_2$/Si capacitors at 300 °C. A systematic increase in peaks of I-

V curves as a function of biasing time (at 075 MV/cm) confirms the diffusion of Cu into Ta_2O_5 at aggressive test conditions.

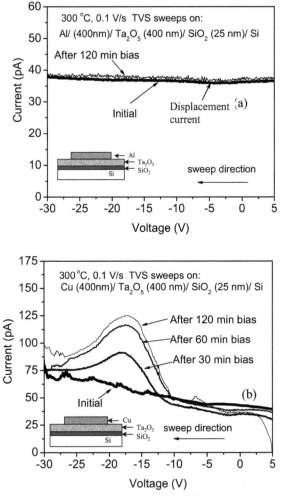

Figure 6.11. TVS results on: (a) Al/Ta₂O₅/SiO₂/Si and (b) Cu/Ta₂O₅/SiO₂/Si capacitors at 300 °C and 0.1 V/s sweep rate. TVS peak grows with biasing time in Cu/Ta₂O₅/SiO₂/Si capacitors.

6.3.3 Mobile Charge (Cu ions) Calculations

The amount of mobile charges is calculated from the flat-band shifts in BTS curves and from the peak area of the TVS curves.

6.3.3.1 Mobile charges using BTS results

The amounts of mobile charges are calculated using Eq. 6.4. In these calculations, it is assumed that singly ionized Cu ions, i.e., Cu^+ are the only type of mobile ions injected and then diffuse/drift into the dielectric. This is assumed because first ionization energy of Cu (7.7 eV/atom) is about one-third of its second ionization energy (20.3 eV/atom), and ions move instantaneously under bias at high temperatures. In addition, solid solubilities of higher ionized states of most metals are much lower than that of singly ionized state [7].

Once Cu^+ are injected, an assumption is to made about the distribution of ions inside the dielectric, i.e., "γ" in Eq. 6.4 is to be defined. In the simplest case, "γ" is 1, when ions pile up at the silicon interface. In the dual dielectric stack, i.e., $Cu/Ta_2O_5/SiO_2/Si$, the question remains as to whether the Cu^+ ions pile up at the Ta_2O_5/SiO_2 interface or at the SiO_2/Si. As concluded in Fig. 6.6, Cu^+ ions (and Al ions) do not diffuse into thermal SiO_2, even under aggressive BTS test conditions. Therefore, it is assumed that the Cu^+ pile up at Ta_2O_5/SiO_2 interface.

The next step is to convert the dual dielectric to an effective Ta_2O_5 capacitor. The thermal SiO_2 thickness is converted into the effective Ta_2O_5 capacitor thickness ($t_{effective\_Ta2O5}$) using

$$\left(\frac{\varepsilon_{SiO_2}}{t_{SiO_2}} \right) = \left(\frac{\varepsilon_{Ta_2O_5}}{t_{effective\_Ta_2O_5}} \right) \tag{6.8}$$

25 nm of thermal SiO_2 used is equivalent to 160 nm $t_{effective\_Ta2O5}$. Now assuming all mobile charge is located at an effective Ta_2O_5 thickness of 160 nm from the Si

interface, "γ" comes out to be 0.71 (= 400/560). Now using Eq. 6.4, Fig. 6.9 is transformed into Fig. 6.12.

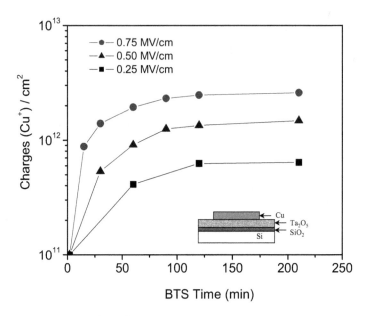

Figure 6.12. Charge/cm^2 (Cu$^+$) diffusion/drift into Ta$_2$O$_5$ as a function of BTS time for different biasing conditions at 300 °C.

6.3.3.1 Mobile charges using TVS results

As discussed in section 6.2.4, the area under the TVS peak gives the estimate of the mobile charges diffusion/drift. Typically, TVS peak areas under-estimate the concentration of the diffused/drifted charges. The charges detected using TVS scans at 300 °C are plotted in Fig. 6.13. The estimated charge concentration is also compared with BTS results under same biasing and temperature conditions. As expected, TVS peak areas give lower values of charge/cm^2 compared to that calculated using flat-band shifts in BTS scans.

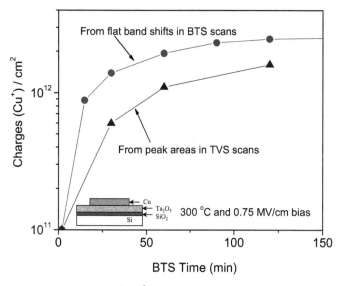

Figure 6.13. **Charge (Cu$^+$)/cm^2 diffusion/drift into Ta$_2$O$_5$ as a function of TVS and BTS time at 0.75 MV/cm bias and 300 °C.**

6.3.4 Measures to Contain Cu-Diffusion

A thin layer of Ti or Ta (~30 nm) is used between Cu and Ta$_2$O$_5$ as a barrier for Cu-diffusion into Ta$_2$O$_5$. Both Ti and Ta are also good adhesion promoters for Cu on Ta$_2$O$_5$ and both can be deposited in the same reactive sputtering system used for Ta$_2$O$_5$ deposition without breaking the vacuum. In addition, being compatible with multilevel metallization processing, both Ti and Ta can readily be integrated at on-chip level.

First, Ti/Ta$_2$O$_5$/SiO$_2$/Si and Ta/Ta$_2$O$_5$/SiO$_2$/Si were fabricated and tested for Ti or Ta diffusion into Ta$_2$O$_5$ at various test conditions. BTS and TVS results on these test structures confirmed that neither Ti nor Ta diffuse into Ta_2O_5, even at aggressive test conditions (0.75 MV/cm and 300 °C). Hence, these samples were used as control samples to determine, if Ti and/or Ta can serve as diffusion barriers for Cu into Ta_2O_5.

After establishing that Ti and Ta do not diffuse into Ta_2O_5, $Cu/Ti/Ta_2O_5/SiO_2/Si$ and $Cu/Ta/Ta_2O_5/SiO_2/Si$ were fabricated and tested for Cu diffusion into Ta_2O_5 through Ti and Ta using BTS and TVS. Any flat-band shift in BTS or growing TVS peak on this capacitor structures will be due to Cu^+ ions (and not from Ti or Ta ions). The BTS and TVS results on Ta_2O_5, $Cu/Ti/Ta_2O_5/SiO_2/Si$ and $Cu/Ta/Ta_2O_5/SiO_2/Si$ capacitors are shown in Fig. 6.14 and 6.15, respectively. No flat-band shift in BTS results and absence of TVS peaks confirm that both Ti and Ta act as excellent diffusion barriers for Cu into Ta_2O_5.

Metal diffusion (or drift) through dielectrics under bias takes place in two steps: (a) first metal atoms ionize to form metal ions at metal/dielectric interface and (b) then metal ions diffuse/drift through the bulk dielectric. For Al, Ti, and Ta, step (b) is the rate-limiting step because all of them form stable oxides (Al_2O_3, TiO_2, Ta_2O_5, respectively) at the metal/Ta_2O_5 interface, and therefore their ions do not drift further into the dielectric. In contrast, Cu does not have a thermodynamically stable oxide and hence, if Cu^+ are available, then they can diffuse through the dielectric. Therefore, formation of Cu^+ ions from Cu can limit the net diffusion/drift process.

In $Cu/Ta_2O_5/SiO_2/Si$ capacitors, Cu is in direct contact with Ta_2O_5. The presence of higher electronegative species, i.e., oxygen atoms, at Cu/Ta_2O_5 interface facilitates the ionization of Cu to Cu^+. In contrast, Ti or Ta layers in $Cu/Ti/Ta_2O_5/SiO_2/Si$ or $Cu/Ta/Ta_2O_5/SiO_2/Si$ capacitors separate Cu from Ta_2O_5 (and hence oxygen atoms). Therefore, the ionization of Cu to Cu^+ is greatly suppressed in $Cu/Ti/Ta_2O_5/SiO_2/Si$ or $Cu/Ta/Ta_2O_5/SiO_2/Si$ capacitors and hence, no Cu ion penetration could be detected using BTS and TVS in $Cu/Ti/Ta_2O_5/SiO_2/Si$ or $Cu/Ta/Ta_2O_5/SiO_2/Si$ capacitors

157

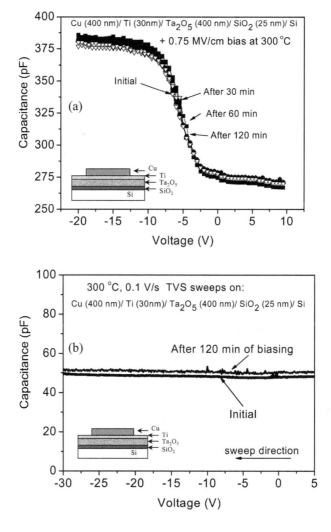

Figure 6.14. (a) BTS and (b) TVS (0.1 V/s sweep) results on Cu/Ti/Ta$_2$O$_5$/SiO$_2$/Si capacitors at 0.75 MV/cm and 300 °C. No mobile Cu ions were detected with BTS or TVS.

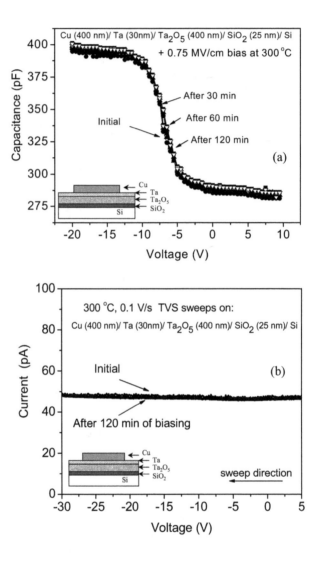

Figure 6.15. (a) BTS and (b) TVS (0.1 V/s sweep) results on Cu/Ta/Ta$_2$O$_5$/SiO$_2$/Si capacitors at 0.75 MV/cm and 300 °C. No mobile Cu ions were detected with BTS or TVS.

6.4   Summary and Conclusions

In this chapter, the extent of Cu diffusion/drift into Ta_2O_5 films is determined and compared with Al, Ta, and Ti at various biasing and temperature conditions using two electrical testing techniques: bias-temperature-stress (BTS) and triangular voltage sweep (TVS). No Cu diffusion was detected at 150 °C at least till 0.75 MV/cm till 120 minutes. However, Cu diffuses/drifts into Ta_2O_5 at aggressive temperature conditions, i.e., at 300 °C. Around 10^{12} charges/cm^2 mobile ions were detected using BTS and TVS. In contrast, no mobile ions were detected with Al, Ti, and Ta as metals in MIS capacitors. It is also demonstrated that a thin layer of Ti or Ta (~30 nm) acts as an excellent Cu diffusion barrier into Ta_2O_5. Hence, Ta_2O_5 decoupling capacitors are compatible with on-chip metallization, and hence suitable for on-chip integration (with Cu as interconnect metallization).

Table 6.1.   Metal ion penetration after 120 minutes of biasing

Conditions	Cu	Al	Ti	Ta	Cu/Ti	Cu/Ta
150 °C and 0.75 MV/cm	none	none	none	none	none	none
300 °C and 0.75 MV/cm	~10^{12} Cu$^+$ ions/cm^2	none	none	none	none	none

6.5   References

6.1   H. B. Bakoglu, *"Circuits, Interconnections, and Packaging for VLSI,"* Addison-Wesley VLSI Systems Series, 1996, ch 7, pp. 281-335.

6.2   R. K. Poon, *"Computer Circuits Electrical Design,"* Prentice Hall, Eaglewood Cliffs, 1995, ch. 8, pp 195-229.

6.3   W. D. Brown, *"Advanced Electronic Packaging with Emphasis on Multichip Modules,"* IEEE Press Series on Microelectronic Systems, 1998, ch. 3, pp. 133-136.

6.4   S. P. Murarka, *"Metallization Theory and Practice for VLSI and ULSI,"* Butterworth-Heinemann, 1993, ch. 12, pp. 196-210.

6.5   S. P. Murarka, Mater. Sci. & Engg. R**19**, 87 (1997).

6.6   S. P. Murarka and S. W. Hymes, Crit. Rev. in Solid State Mater. Sci. **20**, 87 (1995).

6.7   J. D. Mc Brayer, R. M. Swanson, and T. W. Sigmon, J. Electrochem. Soc. **133**, 1241 (1986).

6.8   Y. Shacham- Diamond, A. Dedhia, D. Hoffstetter, and W. G. Oldham, J. Electrochem. Soc. **140**, 2427 (1993).

6.9   J. Pallean, J. C. Oberlin, F. Braud, J. Torres, J. L. Mermet, M. Mouche, A. Ermolief, and J. Piaget, *Proc. Mater. Res. Symp.* 337, 225 (1994).

6.10   G. Raghavan, C. Chiang, P. B. Anders, S. M. Tzeng, R. Villasol, G. Bai, M. Bohr, and D. Fraser, Thin Soild Films **262**, 168 (1995).

6.11   A. L. S. Loke, C. Ryu, C. P. Yue, J. S. H. Cho, and S. S. Wong, IEEE Electron Dev. Lett. **17**, 549 (1996).

6.12   A. Mallikarjunan, Ph.D. thesis, Rensselaer Polytech. Inst., Troy, NY, 1998.

6.13   A. L. S. Loke, J. Wetzel, P. Townsend, T. Tanabe, R. Vrtis, M. Zussman, D. Kumar, C. Ryu, and S. Wong, IEEE Trans. Electron Dev. **46**, 2178 (1999).

6.14   Z. –C. Wu, C. C. Wang, R. G. Wu, J. Electrochem. Soc. **146**, 4290 (1999).

6.15   T. Laurila, K. Zeng, A. Seppala, J. Molarius; I. Suni; and J. K. Kivilahti, *Proc. Mater. Res. Soc. Symposium 2001*, vol. **612**, pp. D7.4 .

6.16   Y. K. Lee, L. K. Maung; K. JaeHyung, T. Osipowicz; and K. Lee, Mater. Sci. & Engg. B 68(2), 99 (1999).

6.17   R. F. Pierret, *"Semiconductor Device Fundamentals,"* Addison Wesley Longman 1996, ch. 16, pp. 563-600.

6.18   T. Suwwan de Felipe, S. P. Murarka, S. Bedell, and W. A. Landford, Thin Solid Films **335**, 49 (1998).

6.19   Y. Shacham-Diamond, A. Dedhia, D. Hoffstetter, and W. G. Oldham, J. Electrochem. Soc.**140**, 2427 (1993).

CHAPTER 7

SUMMARY, CONCLUSIONS, AND REMAINING CHALLENGES

In this chapter, the key results of this thesis are summarized along with the main conclusions. The next set of challenges, which were not investigated in this thesis, are also described.

7.1   Overall Summary of the Thesis

This thesis explores and establishes the advantages of using thin film embedded capacitors as a effective alternative to the currently used surface mount discrete capacitors. Thin film embedded capacitors offer improved electrical performance and higher packaging densities at potentially lower costs of fabrication and placement (or assembly).

This thesis focuses on tantalum oxide (Ta_2O_5) as a dielectric material for embedded thin film capacitors. The prime reason for the choice of Ta_2O_5 is its relatively high dielectric constant (~ 25 for amorphous films compared to ~3.9 for SiO_2 and ~6.9 for Si_3N_4), that is nearly constant through microwave frequencies (up to at least 20 GHz) and its compatibility with established interconnect technologies for IC chips and packages. Pulsed dc reactive magnetron sputtering has been developed and investigated as a method of depositing high quality compound films of desired compositions at low processing temperatures. The following sections summarize key results of this thesis work.

7.1.1   Optimization and Control of Reactive Sputtering Deposition of Ta_2O_5

A major challenge in the pulsed dc reactive sputtering of Ta_2O_5 films is selecting the optimum process conditions to deposit stoichiometric films that exhibit superior and reliable electrical properties while maintaining high deposition rates

163

(discussed in section 3.4). The stoichiometry of the deposited tantalum oxide films, and hence their electrical properties, e.g., breakdown field, leakage current density, and dielectric behavior strongly depend on the oxygen flow rate, sputtering power/current, deposition temperature etc.

In this thesis, an analytical approach is used to systematically evaluate, predict and control the electrical properties of tantalum oxide films as functions of various process parameters. Pulsed dc reactive sputtering process to deposit tantalum oxide films is described using an analytical model. The influences of important process variables, like oxygen flow rate and sputtering ion current, on the steady state oxygen partial pressure in the chamber, deposition rate, as well as film breakdown and leakage characteristics are predicted and analyzed using this model. Experimental results are in close agreement with the model predictions.

This experimentally verified analytical model suggests that during a reactive sputtering process, the steady-state fractional coverages at the target (θ_1), and chamber walls/substrate (θ_2) at a fixed sputtering ion current are determined by the net balance of reaction and sputtering rate. The steady-state fractional coverages at the target (θ_1) and chamber walls/substrate (θ_2) then in turn determine the net deposition rate, film composition and hence the breakdown and leakage characteristics of the deposited films.

The experimentally established existence of multiple oxygen partial pressures at a given oxygen flow rate (hysteresis loop) is theoretically explained using steady state analysis. At very low and high oxygen flow rates, single steady-state oxygen partial pressures exist. At intermediate oxygen flow rates, multiple steady states exist at the same oxygen flow rate, *viz.* lower, intermediate and upper steady states. At an upper steady state, a higher oxygen partial pressure is exhibited, while at a lower steady state, a very low oxygen partial pressure is exhibited.

164

At lower steady states, due to the low oxygen partial pressures, the target is partially oxidized. Thus, the sputtered flux consists of both Ta and TaO_x fragments. The sputtering yield of Ta is an order of magnitude higher than that of oxidized Ta. This causes net higher deposition rate at lower steady states. The presence of metallic tantalum in the films deposited at lower steady states makes them electrically conductive. On the other hand, at upper steady states the target is fully oxidized to tantalum oxide and thus lower deposition rate is exhibited and the deposited films show excellent breakdown and leakage characteristics.

The operating points at upper or lower steady states are stable with respect to any system fluctuations. However, intermediate steady states are unstable to system fluctuations and would toggle to either upper or lower steady state depending upon the nature of the system fluctuation. Hysteresis in partial pressure, deposition rate, breakdown and leakage characteristics during increasing and decreasing flow rate modes is due to the fact that the system follows the lower steady states during increasing flow rate mode and upper steady states during decreasing flow rate mode. The experimental results on our sputtering system indicate that in order to ensure consistently high breakdown fields of tantalum oxide films, oxygen flow rate should be more than the critical flow rate for the increasing flow rate mode.

7.1.2 Breakdown Voltage Versus Capacitance Density and Design Limits

One would like to build capacitors with the highest (known) capacitance density and highest breakdown voltage. However, a trade-off exists between these two properties of interest. This inherent trade-off is explained based on the empirically determined interdependency between breakdown field and dielectric constant of dielectric films.

For highest quality stoichiometric films and homogeneous uniform electric field, the relationship between maximum breakdown field (E_{BR}) and dielectric constant (ε_r) is

165

empirically determined to be E_{BR} (MV/cm) $= \left(20/\sqrt{\varepsilon_r}\right)$. This empirical relationship sets the limit for "best one can achieve" (BOCA) breakdown fields of dielectrics with dielectric constant ε_r.

A design space based on this empirically determined relationship between breakdown field and dielectric constant is developed and explored. The tradeoff between capacitance density and breakdown voltage, with film thickness and dielectric constant as parameters, is established. The validity of the developed design space is verified with tantalum oxide thin films. The breakdown voltages of tantalum oxide films with thickness around 0.15 to 1 μm were in close agreement with the BOCA values. "Thicker" films (>~1 μm) and "thinner" films (< ~0.15 μm) had initially lower breakdown voltages compared to their BOCA values.

Substrate heating during deposition of "thicker" films (> ~1 μm), leads to the deposition of sub-stoichiometric TaO_x, and therefore causes lower breakdown voltages, higher leakage current densities and high dielectric losses. Substrate cooling during deposition ensures the deposition of stoichiometric Ta_2O_5 and restores the breakdown voltages of "thicker" films to their BOCA values.

The lower breakdown fields for "thinner" films (< ~0.15 μm) are due to micro-roughness (peak-to-valley roughness) of the bottom electrode, which can be a significant fraction of the total film-thickness. The micro-roughness of the bottom electrode increases the effective electric field at the sharp and pointed regions of the bottom electrode (also known as 'lightening rod' field amplification), which causes cratering and burning out of the top metal electrode, leading to local and premature breakdown. Using bottom electrode of different P-V roughness, it is shown that in order to achieve BOCA breakdown voltages, "peak-to-valley" roughness at the metal/dielectric interface must be less than 10% of dielectric film thickness.

7.1.2 Broadband Electrical Characterization/Evaluation of Thin Film Capacitors

The advantages offered by the Ta_2O_5 based thin film decoupling capacitors are quantified by evaluating their electrical properties over a wide range of operating frequencies (dc to 20 GHz). High frequency test vehicles were designed and fabricated to evaluate the electrical performance of thin film capacitors till GHz range. The electrical properties evaluated are breakdown field, dielectric constant, leakage current density, dielectric dispersion, and dielectric loss. All the above-mentioned electrical properties are also tested for SiO_2 and Si_3N_4, using same design test vehicles and similar test conditions. Both SiO_2 and Si_3N_4 are standard dielectrics used in microelectronics industry, and hence serve as benchmarks to evaluate the potential of Ta_2O_5 films for decoupling capacitors.

From impedance and insertion loss measurements, the dispersion in dielectric constants of all three materials is evaluated from 1 kHz to 20 GHz. The dielectric constants of SiO_2, Si_3N_4, and Ta_2O_5 exhibit no dispersion up to at least 20 GHz of signal frequency. No dispersions in dielectric constants until 20 GHz indicate that electronic and molecular (atomic) polarization mechanisms are the dominant mechanisms responsible for the dielectric response of these dielectrics. The dielectric losses for all three materials are less than 0.015 over the entire frequency range considered.

The total inductance of the power connect vias is determined from first self-resonance frequencies. The total inductance of power connect vias is determined to be less than 50 pH/μm of via, which is at least two orders of magnitude lower than most discrete capacitors along with connection leads (> 4 nH). Using capacitors with different number of parallel vias, it is demonstrated that several vias in parallel further reduce the overall inductance of the connection paths in thin film capacitors. In addition to low self-inductance, the shorter connections and elimination of connection leads in thin film decoupling capacitors make them highly suitable for integration into 2D and 3D structures. Thin film capacitors based on SiO_2, Si_3N_4, and Ta_2O_5 can provide capacitance densities comparable to the state-of-the-art surface mount discrete decoupling capacitors ($\sim$ 30 nF/cm^2). The leakage current densities below 10^{-6} A/cm^2 (at 0.5 MV/cm) can be achieved with SiO$_2$, Si$_3$N$_4$ and Ta$_2$O$_5$ thin film dielectrics. The breakdown fields and dielectric constants of SiO$_2$, Si$_3$N$_4$ and Ta$_2$O$_5$ closely follow the empirically determined BOCA relationship.

7.1.4 Electrode Metal Diffusion in Tantalum Oxide

On-chip power distribution noise can be suppressed and controlled effectively by integrating (or embedding) thin film decoupling capacitors in the multi-level metallization (MLM) of the ICs. But, selected metals, such as Cu, Al, Ti, and Ta (or their metallic compounds, such as TiN, TaN, etc.) are likely to be used in the MLM layers due to their compatibility with the rest of the BEOL processing steps. Choosing capacitor electrodes to be of same metallization used for on-chip interconnects facilitates the on-chip capacitor integration, due to reduced integration risks. As Cu is becoming the choice of metallization for on-chip level interconnects; the successful on-chip integration of thin film capacitors demands the compatibility of the high-κ dielectrics with Cu.

The extent of Cu diffusion/drift into Ta$_2$O$_5$ films is determined and compared with Al, Ta, and Ti at various biasing and temperature conditions using two electrical testing techniques: bias-temperature-stress (BTS) and triangular voltage sweep (TVS). Cu diffusion was not detected at 150 °C at least up to 0.75 MV/cm, although Cu diffuses/drifts into Ta$_2$O$_5$ at aggressive temperature conditions, i.e., at 300 °C. Approximately $\sim 10^{12}$ charges/cm^2 (mobile ions) were detected using BTS and TVS. In contrast, no mobile ions were detected with Al, Ti, and Ta as metals in MIS capacitors. A thin layer of Ti or Ta ($\sim$30 nm) is shown to be acts an excellent Cu diffusion barrier into Ta$_2$O$_5$.

7.2　Conclusions

The key conclusions derived from the work accomplished in this thesis are listed below.

- Based on empirically determined relationship, E_{BR} (MV/cm) $= \left(20/\sqrt{\varepsilon_r}\right)$, the maximum amount of electrostatic energy that can be stored in any thin film (solid) dielectric material is essentially constant.

   High dielectric constant materials, due to their highly polarizable bond structures, provide a lot of charge trapping centers. A trapped charge creates a very high local electric field in high-κ dielectrics and tends to distort/weaken polar molecular bonds, thereby lowering the enthalpy of activation required for bond breaking. In addition, high-κ dielectrics release more energy per detrapping of trapped charges (to release bond strains) compared to low-κ dielectrics. All these factors combined together cause lower breakdown fields of high-κ dielectrics.

   It is important to mention that this trade-off is valid only for dielectrics in which breakdown is due to the trapping of charge carriers. Hence, for very thin dielectrics (few atoms thick, especially used for gate dielectrics applications), where other breakdown mechanisms, such as tunneling (where charge carriers bypasses the conduction band of the dielectric) are dominant, this relationship does not hold.

- The developed BOCA design space is an effective engineering tool to determine (a) what dielectric constant material and (b) of what film-thickness, is needed to achieve the required capacitance at a given operating voltage. This design space depicts an aggressive trade-off between breakdown field and dielectric constant, and like other designs; a safety margin should be provided to come up to final design value.

- Substrate heating during deposition of "thicker" films (>1 μm) leads to the deposition of sub-stoichiometric tantalum oxide, and hence causes lower breakdown and higher leakage currents. Substrate cooling during deposition ensures the deposition of stoichiometric Ta_2O_5, and restores the breakdown field of "thicker" films to their BOCA values.

- Micro-roughness of the bottom electrode degrades the breakdown fields of thin dielectric films. "Peak-to-Valley" interface roughness must be less than 10% of dielectric film thickness to achieve "best can achieve" breakdown voltages.

- Same area Ta_2O_5 capacitors provide nearly 6 and 3 times better decoupling compared to SiO_2 and Si_3N_4 capacitors, respectively, up to at least 20 GHz frequency. The decoupling performance is proportional to their dielectric constant ratios.

- Ta_2O_5, SiO_2, and Si_3N_4 exhibit no dielectric dispersions up to at least 20 GHz frequencies. Because only electronic and molecular (atomic) polarizations can respond to the ac field in GHz range, these appear to be the dominant mechanisms of polarization in these materials.

- The total inductance of power connect vias in embedded capacitors is less than 50 pH/μm of via, which is at least two orders of magnitude lower than that for most discrete capacitors along with connection leads (> 4 nH). Hence, thin film embedded capacitors offer at least 2 orders of magnitude better performance at GHz frequencies compared to their discrete counterparts. In addition, SiO_2, Si_3N_4, and Ta_2O_5 capacitors can provide capacitance densities comparable to the state-of-the-art surface mount discrete decoupling capacitors (~ 30 nF/cm^2). The leakage current densities exhibited by these thin film capacitors are below 10^{-6} A/cm^2 (at 0.5 MV/cm), which is less than that required in power decoupling capacitors.

- No Cu diffusion was detected at 150 °C at least till 0.75 MV/cm. But Cu diffuses/drifts into Ta_2O_5 at aggressive temperature conditions, i.e., at 300 °C. Thin layer of Ti or Ta (~30 nm) acts as an excellent Cu diffusion barrier into Ta_2O_5. Hence, Ta_2O_5 decoupling capacitors are compatible with on-chip metallization, and hence suitable for on-chip integration (with Cu as interconnect metallization).

- During reactive sputtering of Ta_2O_5, multiple steady state oxygen partial pressures can exist at the same oxygen flow rate. Insulating films are deposited at upper steady states, while metallic films are deposited at lower steady states. Intermediate steady states are metastable. To deposit films with reliable and insulating Ta_2O_5 films, the oxygen flow rates should be kept more than that required to go beyond the hysteresis region in partial pressure versus flow rate curve.

7.3   Next Set of Challenges and Recommendations

- **Interdependency between E_{BR} and ε_r**

The theoretical basis of the empirically determined interdependency between breakdown field and dielectric constant of thin film dielectrics, i.e., E_{BR} (MV/cm) $= \left(20/\sqrt{\varepsilon_r}\right)$, is yet to be fully understood. An understanding based on first principles is desirable to better understand this trade-off and to consider engineered-dielectrics to overcome this trade-off.

- **Test Vehicle Design and Fabrication**

In this work, the performance of thin film decoupling capacitors are evaluated up to 20 GHz. Capacitors fabricated based on the current design resonates (first series resonance) in 10-20 GHz range. Therefore, larger area capacitors cannot be characterized beyond 20 GHz using current high frequency test vehicle.

171

Fabrication of 3D characterization vehicle(s), with better designs, are needed to evaluate the performance of decoupling capacitors up to 200 GHz range. In addition, interaction of signal lines (going through the ground and power planes, shown in Fig. 7.1) with the decoupled power distribution is to be evaluated. The optimization of the fabrication process is required to minimize the masking steps, and to make it cost-effective.

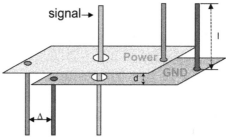

Figure 7.1. **Signal lines going through the power and ground planes of the power distribution system.**

• **Columnar Film Growth**

Cross-sectional SEM images of the tantalum oxide films show columnar film structures (as shown in Fig. 7.2). The fundamental reasons for the columnar growth are yet to be determined. The effect of columnar structure on films electrical properties, such as breakdown field, dielectric constant, leakage current density, dielectric loss and dispersion etc need to be evaluated.

Figure 7.2. **Cross-sectional view of the tantalum oxide film showing the columnar structure.**

172

- **Film Adhesion**

Electrical and mechanical reliability of thin film capacitors strongly depends on the interlayer adhesion. The nature of the interlayer bonds can be physical or chemical. The strength of physical interactions depends on the interface roughness. But excessive interface roughness degrades the breakdown and leakage properties of the dielectric films. In contrast, chemical bonds do not depend on the interface roughness.

The nature of the interface bonds (physical or chemical) need to be evaluated and interlayer adhesion energy is to be determined to evaluate the reliability of multilayer capacitor structures.

- **Novel Capacitor Integration Schemes**

Ta is used as an adhesion promoter and diffusion barrier for Cu metallization into low-κ interlayer dielectrics. Integration schemes should be explored, where these Ta layers (already provided as diffusion barriers and adhesion promoters) can be oxidized to provide dielectric layers. Conceptually, $Ta/Ta_2O_5/Ta$ capacitors can be fabricated at on-chip levels without significantly changing rest of the processing steps.

While this thesis has extensively addressed and explored several key issues and challenges in developing and integrating thin film capacitors, many more reliability and tolerance issues will be involved in putting this work into packaged electronic products.

APPENDIX A

SURFACE ROUGHNESS GROWTH IN REACTIVELY
SPUTTERED Ta₂O₅

A.1. Introduction

In this thesis, reactively sputtered Ta_2O_5 is explored as the dielectric for embedded capacitors for advanced packaging applications. Besides that, Ta_2O_5 has also been considered a promising high permittivity (high-k) dielectric material for DRAM cells [1-3] and alternative gate dielectric material for silicon dioxide [4]. In all these applications, the micro-roughness at metal-dielectric interfaces significantly degrades the leakage and breakdown characteristics of thin film capacitors [5].

The growth front roughness of Ta_2O_5 amorphous films deposited by pulsed dc reactive sputtering is investigated using atomic force microscopy, and is discussed in this appendix. Film deposition during reactive sputtering is explained based on dynamic scaling hypothesis in which both time and space scaling are considered simultaneously. The scaling exponents are extracted, and a re-emission model is suggested to describe the possible surface growth mechanisms during reactive sputtering. Monte Carlo simulations were carried out based on a recently developed re-emission model [6,7], where incident flux distribution, shadowing, sticking coefficient, and surface diffusion mechanisms were accounted for in the deposition process.

A.2 Experimental Methodology

Ta_2O_5 thin films of various thicknesses (0.14 to 5.8 μm) were deposited by pulsed plasma magnetron dc reactive sputtering. Film thickness was measured after the film deposition using a variable-angle spectroscopic ellipsometer (VASE®) and also verified by step-height measurements using profilometer. The surface morphology was measured using contact-mode AFM (atomic force microscopy) (Park Scientific Auto CP). The radius of the Si_3N_4 tip was about 10 nm, and the side angle is about 10°. We took 8 AFM images, of size 4μm×4μm per sample and each image included 256 x 256 pixels. PSI ProScan software was used to extract height-height correlation function from the AFM scans of different thickness of Ta_2O_5 films.

A.3   Results and Discussion

The quantitative information of the surface morphology is extracted from the equal-time height-height correlation function H(r, t), defined as $H(r) = < [h(r)-h(0)]^2 >$. Here h(r) is the surface height at position r (x, y) on the surface relative to the mean surface height. The notation <...> means a statistical average. The dynamic scaling hypothesis requires that $H(r) \sim r^{2\alpha}$ for r<<ξ, and $H(r)=2w^2$ for r>>ξ. Here ξ is the lateral correlation length, w is the interface width or rms roughness, and α is the roughness exponent (local), which describes the surface fractality.

The interface width w increases as a power law of deposition time t, $w \sim t^\beta$, where β is the growth exponent, and the lateral correlation length ξ grows as $\xi \sim t^{1/z}$, where 1/z is the dynamic exponent. Dynamic scaling also requires that $z = (\alpha/\beta)$. Therefore, from the slopes of linear fits to the log-log plots of H(r) versus r (for r << ξ), w versus t, and ξ versus t, we can extract the roughness exponents α, β, and 1/z respectively.

Representative surface morphologies of reactively sputtered Ta_2O_5 films (from AFM scans) are shown in Fig. A.1(a), for the deposition times of t = 3, 15, 48, and 120 minutes. Fig. A.1(a) clearly shows that the surface features grow with time. Height-height correlation function curves, H(r,t), are calculated from AFM images after a line-by-line averaging process. The height-height correlation function curves are plotted on a log-log scale in Fig. A.1(b), for different deposition times from 3 to 120 minutes.

The roughness exponent (α) is calculated from the slope ($=2\alpha$) of the height-height correlation curve on a log-log plot for $r \ll \xi$. The interface width w is obtained from the root mean square fluctuations of the height data. In order to determine ξ accurately, the two-dimensional auto-correlation $C(r) = \langle h(r)\ h(0) \rangle$ function is calculated from each AFM image, and the quadrant circularly-averaged auto-correlation function $C_c(r)$ is used to determine ξ by the relation $C_c(\xi)=C_c(0)/e$.

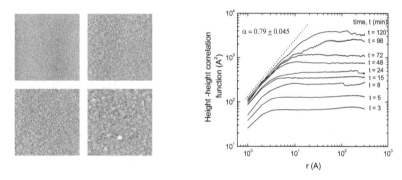

Figure A.1. **(a) Representative surface images of amorphous Ta$_2$O$_5$ films measured by AFM, for growth times of t = 3, 15, 48, and 120 minutes. (b) The equal-time height-height correlation function H(r, t) as a function of the distance r is plotted in log-log scale. All H(r, t) curves gives the same α value within the short-range spatial scaling regime as indicated by the dashed line.**

The roughness exponent (α) is calculated from the slope ($=2\alpha$) of the height-height correlation curve on a log-log plot for $r \ll \xi$. The interface width w is obtained from the root mean square fluctuations of the height data. In order to determine ξ accurately, the two-dimensional auto-correlation $C(r) = \langle h(r)\ h(0) \rangle$ function is calculated from each AFM image, and the quadrant circularly-averaged auto-correlation function $C_c(r)$ is used to determine ξ by the relation $C_c(\xi)=C_c(0)/e$.

In Fig. A.1(b), the interface width and the lateral correlation length grow with time, but the slope of the height-height correlation curve for $r \ll \xi$. ($=2\alpha$) is nearly constant. For initial phase of deposition (t < 10 minutes), the height-height correlation curves do not overlap each other for r<<ξ. In the literature this type of growth is termed non-stationary growth [8]. In non–stationary growth, the fluctuations and the smoothening effects cannot quite reach a balance and stationary local slope cannot be achieved. In a non-stationary growth the scaling relation $z = (\alpha/\beta)$ may breakdown [9]. After around 10 minutes, the height-height correlation curves begin to overlap, which suggests that the growth gradually becomes stationary.

The roughness exponent value of each H(r, t) curve is the same within the experimental errors: $\alpha = 0.79 \pm 0.04$. But measured α values can be higher than the true values because of the tip effect [10, 11]. Aue et al. [10] showed that the surface fractal dimension, determined with a scanning probe technique, will always lead to an underestimate of the actual scaling dimension, due to the convolution of tip and surface (fractal dimension d_f for 1+1 interface is related to α by $d_f = 2-\alpha$ [12]). Aue et al. [10] analysis included tips with different shapes and aspect ratios. Their analysis for a tip similar to what we used suggests that after AFM tip adjustments, the value of α for our results will be around 0.6-0.7.

In Fig. A.2(a), the interface width w is plotted as a function of deposition time t. The linear fit to Fig. A.2(a) gives the growth exponent $\beta = 0.45 \pm 0.03$. In this fit, data points up to t = 8 minutes are not included, because the growth is non-stationary till around 10 minutes. The log-log plot of ξ versus t is shown in Fig. A.2(b). As evident from Fig. A.2(b), ξ does not change too much up to the deposition times around 10 minutes, after which a stationary growth begins to develop. After around 10 minutes, ξ increases linearly with deposition time on log-log plot, and the best fit to this regime gives $1/z = 0.61 \pm 0.07$. From the measured α and β, the scaling relation $z = (\alpha/\beta)$, predicts $1/z = 0.56$ ($z = 1.75$), which is close to the measured $1/z$ value. Our experimental growth exponent value of $\beta = 0.45 \pm 0.03$ is close to the reported values of $\beta \sim 0.4$ in the literature from the elemental sputtering depositions [13].

Our experimental results show that the roughness exponents obtained for our sputter deposited films fit none of the presently known universality classes [12,14].

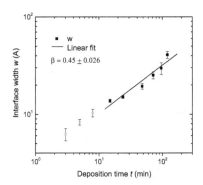 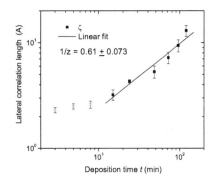

Figure A.2. **(a) The interface width w versus growth time t is plotted in log-log scale. The best linear fit gives the growth exponent β. (b) The lateral correlation length ξ versus growth time t is plotted in log-log scale. The best linear fit gives the dynamic exponent 1/z in log-log scale.**

local models typically give $\beta \leq 0.25$ except the random deposition model where $\beta = 0.50$. In a local growth model, each surface point is related only to a limited number of neighboring points. In reality, the morphology of the neighboring points can result in a shadowing effect where the valleys of the surface receive less incident flux due to the hills around them. A pure shadowing effect would give $\beta = 1$ [7,15,16]. Furthermore, an incident atom can bounce off of the surface depending on its sticking coefficient. The average probability of sticking is a result of the complicated interactions between the incident atom and the surface. These re-emitted particles can fill the valleys faster and increase the conformality. Another smoothening effect comes from adatom diffusion on the surface towards regions with lower surface potential energy (e.g. valleys), which is a local effect.

Therefore, Monte Carlo simulations were carried out based on the recently developed re-emission model by Drotar et al. [6,7] and Karabacak et al. [3,17], where incident flux distribution, sticking coefficient, and surface diffusion were accounted for in the deposition process during reactive sputter deposition. The details of the basic processes used in the simulation code are given in references [6,7].

A summary of basic deposition processes is sketched in Fig. A.3(a). A single particle (with a position described by x, y, and z) is introduced with random (uniformly distributed) variables x, and y, while z is set to the maximum height of the surface, plus one. The direction of the particle follows the distribution $dP(\theta,\phi)/d\Omega = \cos\theta/(2\pi \sin \theta)$, where ϕ is the angle of the projection of the particle's trajectory in the x-y plane, θ is the angle between the particle's trajectory and the negative z axis, and $d\Omega$ is equal to $d(\cos\theta)d\phi$. The particle moves in a straight line until it hits the surface and it is either deposited (h→h+1) or is re-emitted according to the thermal re-emission mode.

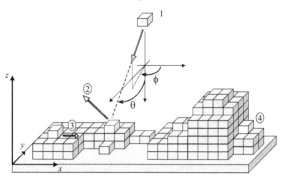

Figure A.3.   A schematic of Monte Carlo simulation

The particle is allowed to continue 'bouncing' off the surface until it is deposited on the surface or heads away from the surface. Sticking probabilities of the first impact (zero[th] order) and of following impacts (n[th] order) of the re-emitted particle are defined by s_0 and s_n, respectively. Once a particle is deposited, a prescribed number of atoms, being set to D/F, are randomly picked to become candidates for diffusion. Here, D denotes the number of surface atoms that are available to diffuse within the unit time interval, in which F atoms are deposited to the surface. The particle goes on jumping until it finds an island of atoms, a kink site, a valley or any lattice point, where the diffusion probability becomes small. Finally, after the diffusion step is done, another particle is allowed to fall on the surface and the whole process is repeated.

During our simulations we studied the change of β for various values of sticking coefficients (with $s_0 \leq 1$ and $s_{n>0} = 1$) and surface diffusion rates. We observe that high sticking coefficients values ($s_0 \sim 0.8 - 1$) give $\beta = 1$ for all the diffusion rates studied ($0 \leq D/F \leq 100$). This result is expected due to the shadowing effects from the incident flux. Therefore, our experimental results cannot be explained if we assume sticking coefficients close to unity. We also studied the low sticking coefficients ($s_0 \sim 0.1 - 0.4$) and they give $\beta \sim 0$, which is much lower then what we observe experimentally. However, for intermediate values of s_0 we obtain β ranging from ~ 0 to 1. We also see that for sticking coefficients around $s_0 \sim 0.60$-0.65 growth exponent gets values $\beta \sim 0.3$-0.5 (for various D/F), which is close to our experimental $\beta \sim 0.45$ and to $\beta \sim 0.4$ from the elemental material sputtering depositions reported in the literature.

A.4. Conclusions

The dynamic growth front roughening of Ta_2O_5 amorphous films grown by pulsed plasma d.c. reactive sputtering can be explained based on dynamic scaling hypothesis, in which both time and space scaling are considered. The interface width w increases as a power law with deposition time t, $w \sim t^{\beta}$, with $\beta = 0.45 \pm 0.03$. The lateral correlation length ξ grows as $\xi \sim t^{1/z}$, with $1/z = 0.61 \pm 0.07$. The roughness exponent extracted from the slope of height-height correlation analysis is $\alpha = 0.79 \pm 0.04$. The results are similar to that obtained by sputtering of elemental materials, and they don't fit any of the presently known growth models. However, recently developed re-emission model can bring out a good degree of explanation of the surface dynamics of our sputter deposited films. The model includes sticking coefficient, re-emission (of the non-sticking particles), shadowing, and surface diffusion processes. We discuss that the sticking coefficients of the incident particles should be less than unity in order to explain our experimental scaling exponents.

A.5. References

A.1.   C. Hashimoto, H. Oikawa, and N. Honma, IEEE Trans. Electron Devices **36**, 14 (1989).

A.2.   G. Q. Lo, D. L. Kwong, P. C. Fazan, V. K. Mathews, and N. Sandler, IEEE Electron Device Lett. **14**, 216 (1993).

A.3.   R. Ulrich and L. Schaper, *Proc. International Symp. on Advanced Packaging Materials* 2000, pp. 38-43.

A.4.   C. Chaneliere, J. L. Autran, R. A. B. Devine, and B. Balland, Mater. Sci. Eng. **R. 22**, 269 (1998).

A.5.   Y.–P. Zhao, G.-C Wang, T.-M Lu, G. Palasantas, and J. Th. M. De Hosson, Phys. Rev. B**60**, 9157 (1999).

A.6.   J. T. Drotar, Y.–P. Zhao, T.-M Lu, and G.-C Wang, Phys. Rev. B**61**, 3012 (2000).

A.7.   A.6.7.J.T. Drotar, Y.–P. Zhao, T.-M Lu, and G.-C Wang, Phys. Rev. B**62**, 2118 (2000).

A.8.   T.-M. Lu, H.-N. Yang, and G.-C. Wang, in Fractal Aspects of Materials, edited by F. Family, P. Meakin, B. Sapoval, and R. Wool, Mat. Res. Soc. Symp. Proc. **367**, 283 (1995).

A.9.   Y.-P. Zhao, G.-C. Wang, and T.-M. Lu, Characterization of Amorphous and Crystalline Rough Surfaces: Principles and Applications, (Academic Press, San Diego, 2000).

A.10.   J. Aue and J. Th. M. De Hosson, Appl. Phys. Lett. **71**, 1347 (1997).

A.11.   Q.M. Hudspeth, K.P. Nangle, Y.-P. Zhao, T. Karabacak, , C.V. Nguyen, M. Meyyappan, G.-C. Wang, and T.-M. Lu, Surface Science, **515/2-3**, 453-461(2002).

A.12.   A.-L. Barabasi and H. E. Stanley, Fractal Concepts in Surface Growth (Cambridge University, Cambridge, England, 1995.

A.13.   See Refs. 3 and H. You, R. P. Chiarello, H. K. Kim, and K. G. Vadervoort, Phys. Rev. Lett **70**, 2900 (1993); A. E. Lita and J. E. Sanchez, Phys. Rev. B**61**, 7692 (2000); A. E. Lita and J. E. Sanchez, J. Appl. Phys. **85**, 876 (1999).

A.14.   F. Family and T. Vicsek, Dynamics of Fractal Surfaces (World Scientific, Singapore, 1991).

A.15.   R. P. U. Karunasiri, R. Bruinsma, and J. Rudnick, Phys. Rev. Lett. 62, 788 (1989).

A.16.   J.-H. Yao and H. Guo, Phys. Rev. E 47, 1007 (1993).

A.17.   T. Karabacak, Y.-P. Zhao, G.-C. Wang, T.-M. Lu, Phys. Rev. B**66**, 075329 (2002).

APPENDIX B

PARAMETERS EXTRACTION FROM S_{21} MEASUREMENTS

Two-port measurement of S_{21} is made on the capacitors using network analyzer. During the measurement, one probe is connected to the signal generator and the others are connected to the terminating resistor, both through 50 Ω transmission lines. The probe has a GSG (Ground-Signal-Ground) configuration.

The following discussion describes the equations used to determine the capacitance of the dielectric material from S_{21} parameters, which were measured using the HP8510C Vector Network Analyzer. S_{21} parameters are obtained from the network analyzer in the form of magnitude and phase angle. First they are converted to the A+ jB form and their real and imaginary parts will be used for the extraction of the capacitance as described below.

B.1. Series Model for the Capacitor

In this case, the capacitor is modeled as a capacitive element in series with a resistance. The S parameters can be derived from the ABCD parameters by considering the circuit shown below.

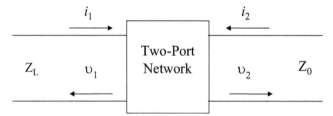

Figure. B.1.   Schematic diagram of the two-port network.

$$v_1 = Av_2 + Bi_2 \tag{B1}$$

$$i_1 = Cv_2 + Di_2 \tag{B2}$$

The reflection coefficient, Γ is given by

$$S_{11} = \Gamma = \Gamma_r + j\Gamma_i = \frac{Z_L - Z_0}{Z_L + Z_0} \tag{B3}$$

From Fig. B.2, we have,

$$\frac{v_2}{i_2} = Z_0 \text{ and } \frac{v_1}{i_1} = Z_L \tag{B4}$$

Using equations (B1), (B2) and (B4), Z_L can be written as

$$Z_L = \frac{v_1}{i_1} = \frac{Av_2 + Bi_2}{Cv_2 + Di_2} = \frac{Ai_2 Z_0 + Bi_2}{Ci_2 Z_0 + Di_2} = \frac{AZ_0 + B}{CZ_0 + D} \tag{B5}$$

Inserting equation (B5) into (B6) results in the following expression,

$$S_{11} = \Gamma = \frac{AZ_0 + B - Z_0(CZ_0 + D)}{AZ_0 + B + Z_0(CZ_0 + D)} = \frac{A + BY_0 - CZ_0 - D}{A + BY_0 + CZ_0 + D} \tag{B6}$$

The whole S parameter matrix can be obtained from the ABCD parameters by the same procedure.

$$[S] = \begin{bmatrix} S_{11} & S_{12} \\ S_{21} & S_{22} \end{bmatrix} = \begin{bmatrix} \dfrac{A + BY_0 - CZ_0 - D}{A + BY_0 + CZ_0 + D} & \dfrac{2(AD - BC)}{A + BY_0 + CZ_0 + D} \\ \dfrac{2}{A + BY_0 + CZ_0 + D} & \dfrac{-A + BY_0 - CZ_0 + D}{A + BY_0 + CZ_0 + D} \end{bmatrix} \tag{B7}$$

For the shunt admittance, the ABCD matrix is given as,

$$\begin{bmatrix} V_1 \\ I_1 \end{bmatrix} = \begin{bmatrix} A & B \\ C & D \end{bmatrix} \begin{bmatrix} V_2 \\ I_2 \end{bmatrix} \tag{B8}$$

where,

$$A = \frac{V_1}{V_2}\bigg|_{I_2=0} = 1 \qquad\qquad B = \frac{V_1}{I_2}\bigg|_{V_2=0} = 0$$

184

$$C = \frac{I_1}{V_2}\bigg|_{I_2=0} = G + jB = Y_L = \frac{1}{Z_L} \qquad\qquad D = \frac{I_1}{I_2}\bigg|_{V_2=0} = 1 \qquad\qquad \text{(B9)}$$

The ABCD matrix for shunt admittance then becomes,

$$\begin{bmatrix} A & B \\ C & D \end{bmatrix} = \begin{bmatrix} 1 & 0 \\ Y_L & 1 \end{bmatrix} \qquad\qquad \text{(B10)}$$

Inserting the constants from equation (B10) into (B7) we get,

$$S_{11} = \frac{-Y_L Z_0}{2 + Y_L Z_0} = \frac{-Z_0}{2Z_L + Z_0} \qquad\qquad \text{(B11)}$$

and

$$S_{21} = \frac{2}{2 + Y_L Z_0} = \frac{2Z_L}{2Z_L + Z_0} \qquad\qquad \text{(B12)}$$

Arranging the right hand side of the equation (B12) using $S_{21} = S_{21r} + jS_{21i}$ we get,

$$Z_L = \frac{Z_0}{2}\left[\frac{(-S_{21r}^2 - S_{21i}^2 + S_{21r}) + jS_{21i}}{(1 - S_{21r})^2 + S_{21i}^2}\right] \qquad\qquad \text{(B13)}$$

Comparing equation (B13) with $Z_L = R_L + jX_L$, we get

$$R_L = \frac{Z_0}{2}\left[\frac{-S_{21r}^2 - S_{21i}^2 + S_{21r}}{(1 - S_{21r})^2 + S_{21i}^2}\right] \qquad\qquad \text{(B14)}$$

and

$$X_L = \frac{Z_0}{2}\left[\frac{S_{21i}}{(1 - S_{21r})^2 + S_{21i}^2}\right] \qquad\qquad \text{(B15)}$$

For a capacitor,

$$Z_L = R_L + jX_L = R_L + \frac{1}{j\omega C} \qquad\qquad \text{(B16)}$$

From equations (B15) and (B16) we can get the expression for capacitance,

$$C = -\frac{1}{X_L \omega} = -\frac{2}{Z_0 \omega}\left[\frac{(1 - S_{21r})^2 + S_{21i}^2}{S_{21i}}\right] \qquad (B17)$$

B.2. Parallel Model of the Capacitor

In a parallel model, Y_L is the admittance and G_L and B_L are the conductance and susceptance respectively.

From Equation (B12), S_{21} can be expressed as,

$$S_{21} = \frac{2}{2 + Y_L Z_0} \qquad (B18)$$

Substituting $S_{21} = S_{21r} + jS_{21i}$ in the above equation and rearranging gives,

$$Y_L = \frac{2}{Z_0}\left[\frac{(-S_{21r}^2 - S_{21i}^2 + S_{21r}) - jS_{21i}}{S_{21r}^2 + S_{21i}^2}\right] \qquad (B19)$$

Using, $Y_L = G_L + jB_L$,

$$G_L = \frac{2}{Z_0}\left[\frac{-S_{21r}^2 - S_{21i}^2 + S_{21r}}{S_{21r}^2 + S_{21i}^2}\right] \qquad (B20)$$

and

$$B_L = \frac{2}{Z_0}\left[\frac{-S_{21i}}{S_{21r}^2 + S_{21i}^2}\right] \qquad (B21)$$

The capacitance is calculated as follows,

$$C = \frac{B_L}{\omega} = \frac{-2}{\omega Z_0}\left[\frac{S_{21i}}{S_{21r}^2 + S_{21i}^2}\right] \qquad (B22)$$

B.3. Insertion Loss

Rather than S_{21}, insertion loss (attenuation) is an important quantity for comparing the performance of two dielectrics. By looking at the insertion loss, it is possible to get a qualitative picture of the performance of the dielectric at high frequency.

Insertion loss is calculated as,

$$\text{Insertion Loss} = -20\log|S_{21}| \tag{B23}$$

From equation (B18),

$$\left|\frac{1}{S_{21}}\right| = \sqrt{\left(1 + \frac{G}{2Y_0}\right)^2 + \left(\frac{B}{2Y_0}\right)^2} \tag{B24}$$

Hence,

$$\text{Insertion Loss} = 20\log\left|\frac{1}{S_{21}}\right| = 10\log\left[\left(1 + \frac{G}{2Y_0}\right)^2 + \left(\frac{B}{2Y_0}\right)^2\right] \tag{B25}$$

Assuming a very good dielectric material with little conductance, i.e. G<<B, Insertion loss can be approximated as,

$$\text{Insertion Loss} \cong 10\log\left[\left(\frac{B}{2Y_0}\right)^2\right] \tag{B26}$$

VDM

Verlag
Dr. Müller

Wissenschaftlicher Buchverlag bietet

kostenfreie

Publikation

von

wissenschaftlichen Arbeiten

Diplomarbeiten, Magisterarbeiten, Master und Bachelor Theses
sowie Dissertationen, Habilitationen und wissenschaftliche Monographien

Sie verfügen über eine wissenschaftliche Abschlußarbeit zu aktuellen oder zeitlosen
Fragestellungen, die hohen inhaltlichen und formalen Ansprüchen genügt,
und haben **Interesse an einer honorarvergüteten Publikation**?

Dann senden Sie bitte erste Informationen über Ihre Arbeit per Email
an info@vdm-verlag.de. Unser Außenlektorat meldet sich umgehend bei Ihnen.

VDM Verlag Dr. Müller Aktiengesellschaft & Co. KG
Dudweiler Landstraße 125a
D - 66123 Saarbrücken

www.vdm-verlag.de